T0269050

These lecture notes are intended as a non-technical overview of scattering theory. The point of view adopted throughout is that scattering theory provides a parametrization of the continuous spectrum of an elliptic operator on a complete manifold with uniform structure at infinity. The simple and fundamental case of the Laplacian on Euclidean space is described in the first two lectures to introduce the basic framework of scattering theory. In the next three lectures various results on Euclidean scattering, and the methods used to prove them, are outlined. In the last three lectures these ideas are extended to non-Euclidean settings.

These lecture notes will be of interest to researchers and graduate students in analysis and differential geometry.

Geometric scattering theory

Geometric scattering theory

RICHARD B. MELROSE

Massachusetts Institute of Technology

CAMBRIDGE
UNIVERSITY PRESS

CAMBRIDGE UNIVERSITY PRESS
Cambridge, New York, Melbourne, Madrid, Cape Town, Singapore,
São Paulo, Delhi, Dubai, Tokyo, Mexico City

Cambridge University Press
The Edinburgh Building, Cambridge CB2 8RU, UK

Published in the United States of America by Cambridge University Press, New York

www.cambridge.org
Information on this title: www.cambridge.org/9780521496735

First published 1995

A catalogue record for this publication is available from the British Library

ISBN 978-0-521-49673-5 Hardback
ISBN 978-0-521-49810-4 Paperback

Contents

List of Illustrations

Preface

These notes are based on lectures delivered at Stanford University in January[1] 1994 and then repeated at MIT in the Spring semester. I am very grateful to the members of the Mathematics Department at Stanford, and in particular Ralph Cohen, for the invitation and hospitality. My especial thanks to those who attended the lectures and contributed in one way or another. I am particularly pleased to acknowledge the influence on my thinking of two of the members of the audience, Ralph Phillips and Joe Keller. Rafe Mazzeo encouraged me to write up the lectures, provided me with his own notes and, as if that were not enough, made many helpful comments on the manuscript. I should also like to extend my thanks to Sang Chin, Daniel Grieser, Andrew Hassell, Mark Joshi, Olivier Lafitte, Eckhard Meinrenken, Edith Mooers and Andras Vasy, who attended the second hearing[2] of the lectures at MIT and together made many useful remarks; Andras Vasy was particularly helpful in reading and correcting the notes as they dribbled out. I would also like to thank Tanya Christiansen and Gunther Uhlmann for their assistance and Lars Hörmander, Georgi Vodev and Maciej Zworski for their comments on later versions of the manuscript.[3]

It is my hope that these notes may serve as an introduction to an active and growing area or research, although I fear they represent a rather steep learning curve.

---

[1] It was a horrible month in Cambridge I am told, very pleasant indeed in Palo Alto. This footnote is an indication of things to come in the body of the notes. If you can't stand it, stop now!

[2] Of course I had really wanted to do things in the other order but did not manage to get my thoughts together in time.

[3] Of course, I claim sole credit for all remaining errors.

Introduction

The lectures on which these notes are based were intended as an, essentially non-technical, overview of scattering theory. The point of view adopted throughout is that scattering theory provides a parametrization of the continuous spectrum of an elliptic operator on a complete manifold with uniform structure at infinity. The simple, and fundamental, case of the Laplacian on Euclidean space is described in the first two lectures to introduce the basic framework of scattering theory. In the next three lectures various results on Euclidean scattering, and the methods used to prove them, are outlined. In the last three lectures these ideas are extended to non-Euclidean settings. This is an area of much current research and my idea was to show how similar the Euclidian and the less familiar cases are. Some of the interactions of scattering theory with hyperbolic geometry, index theory and Hodge theory are also indicated.

I have made no attempt at completeness here but simply described what time, and my own tastes, indicate. In particular there should be at least three times as many references as there are. If I have offended by omitting reference to important work, this should not be interpreted as a deliberate slight! In writing up the lectures I have made extensive use of footnotes to cover more subtle points, to clarify statements that were felt to be obscure, by someone, and to make comments. These asides can be freely ignored.

1
Euclidean Laplacian

1.1 The Laplacian

A fundamental aspect of scattering theory, and one to which I shall give considerable emphasis, is the parametrization of the continuous spectrum of differential operators, especially the Laplace operator. I therefore want to begin these lectures with a discussion of the spectral theory of the flat Laplacian on Euclidean space:

$$(1.1) \qquad \Delta = D_1^2 + D_2^2 + \cdots + D_n^2 \text{ on } \mathbb{R}^n, \ D_j = \frac{1}{i} \frac{\partial}{\partial z_j}$$

where $z_1, \ldots, z_n$ are the standard coordinates. Notice that this is the 'geometer's[1] Laplacian' whereas the 'analyst's Laplacian' $-\Delta$.[2]

To a large extent below, except where it is really important, I shall avoid functional analytic statements relating to the boundedness of operators on Hilbert spaces. Thus I shall consider, at least initially, Δ as an operator on Schwartz' space[3] of $\mathcal{C}^\infty$ functions which decrease rapidly at infinity with all derivatives:

$$(1.2) \qquad \mathcal{S}(\mathbb{R}^n) = \left\{ u : \mathbb{R}^n \longrightarrow \mathbb{C}; \ \sup_{z \in \mathbb{R}^n} |z^\alpha D^\beta u(z)| < \infty \right\}.$$

---

[1] Of course it depends on the sort of 'geometer' you know; this positive Laplacian is the 0-form case of the Hodge Laplacian. Some geometers use the analysts' convention.

[2] The 'scattering theorist's Laplacian' is either $-i\Delta$ or $A = \begin{pmatrix} 0 & \mathrm{Id} \\ \Delta & 0 \end{pmatrix}$. The reason for considering A should become clearer in Section 3.2.

[3] See [42], Definition 7.1.2. It is somewhat contradictory to be using $\mathcal{S}(\mathbb{R}^n)$, which is a more subtle space topologically than are Hilbert spaces such as $L^2(\mathbb{R}^n)$; nevertheless doing so avoids the discussion of unbounded operators. See also [103].

The Fourier transform

$$(1.3) \qquad \widehat{f}(\zeta) = \int_{\mathbb{R}^n} e^{-iz\cdot\zeta} f(z)dz$$

is an endomorphism[4] of $\mathcal{S}(\mathbb{R}^n)$ with inverse

$$(1.4) \qquad f(z) = (2\pi)^{-n} \int_{\mathbb{R}^n} e^{iz\cdot\zeta} \widehat{f}(\zeta)d\zeta.$$

Since[5] $\widehat{D_j f} = \zeta_j \widehat{f}$, conjugation by the Fourier transform reduces any constant coefficient operator to multiplication by a function, in particular

$$(1.5) \qquad \widehat{\Delta f} = |\zeta|^2 \widehat{f}, \ \forall \ f \in \mathcal{S}(\mathbb{R}^n).$$

1.2 Spectral resolution

Using (1.5), and the inversion formula (1.4), the form of the spectral resolution [6] of Δ can be readily deduced. Introducing polar coordinates, $\zeta = \lambda\omega$, $\lambda = |\zeta|$ in (1.4) gives

$$(1.6) \qquad f(z) = (2\pi)^{-n} \int_0^\infty \int_{\mathbb{S}^{n-1}} e^{i\lambda z\cdot\omega} \lambda^{n-1} \widehat{f}(\lambda\omega)d\omega d\lambda.$$

This can be rewritten as a decomposition of the identity operator:

$$(1.7)$$

$$\mathrm{Id} = \int_0^\infty E_0(\lambda)d\lambda, \quad E_0(\lambda)f = (2\pi)^{-n} \int_{\mathbb{S}^{n-1}} e^{i\lambda z\cdot\omega} \lambda^{n-1} \widehat{f}(\lambda\omega)d\omega.$$

---

[4] See [42], Theorem 7.1.5.
[5] See [42], Lemma 7.1.4.
[6] See [98] for a discussion of the spectral theorem; it is not necessary to know this result to proceed (in fact this admonition could be appended to many subsequent comments).

Here[7] $E_0(\lambda)d\lambda$ is a projection-valued measure[8] which gives the spectral decomposition of the Laplacian

$$(1.8) \qquad \Delta = \int_0^\infty \lambda^2 E_0(\lambda)d\lambda.$$

The operator $E_0(\lambda)$ has range in the null space of $\Delta - \lambda^2$; as follows from the fact that the 'plane waves' $\Phi_0(z, \omega, \lambda) = \exp(i\lambda z \cdot \omega)$ are, for $\omega \in \mathbb{S}^{n-1}$, solutions of $(\Delta - \lambda^2)\Phi_0 = 0$. It is convenient to decompose $E_0(\lambda)$ as a product of two operators. Define[9]

$$(1.9) \quad (\Phi_0(\lambda)g)(z) = \int_{\mathbb{S}^{n-1}} \Phi_0(z, \omega, \lambda)g(\omega)d\omega, \ \ \Phi_0(z, \omega, \lambda) = e^{i\lambda z \cdot \omega}.$$

Thus $\Phi_0(\lambda) : \mathcal{C}^\infty(\mathbb{S}^{n-1}) \longrightarrow \mathcal{S}'(\mathbb{R}^n)$, the space of tempered distributions. The formal adjoint operator is just[10]

$$(1.10)$$

$$(\Phi_0^*(\lambda)f)(\omega) = \int_{\mathbb{R}^n} \Phi_0(z, \omega, -\lambda)f(z)dz, \ \ \Phi_0^*(\lambda) : \mathcal{S}(\mathbb{R}^n) \longrightarrow \mathcal{C}^\infty(\mathbb{S}^{n-1}),$$

since $\Phi_0(z, \omega, -\lambda) = \overline{\Phi_0(z, \omega, \lambda)}$. Then the definition (1.7) becomes

$$(1.11) \qquad E_0(\lambda) = (2\pi)^{-n}\lambda^{n-1}\Phi_0(\lambda)\Phi_0^*(\lambda), \ \lambda > 0.$$

Now, for fixed $0 \neq \lambda \in \mathbb{R}$, $\Phi_0^*(\lambda)$ is surjective as a map (1.10).[11] Thus to compute the range of $E_0(\lambda)$ it is only necessary to find the range of $\Phi_0(\lambda)$. In fact it is as large as could reasonably be expected.

---

[7] λ is the 'frequency' of the wave $e^{i\lambda z \cdot \omega}$.

[8]   It is not the case that $E_0(\lambda)$ maps $f \in \mathcal{S}(\mathbb{R}^n)$ into $\mathcal{S}(\mathbb{R}^n)$; the form of the range is discussed below. One might therefore wonder on what space this is supposed to be a projection! One way to explain this is in terms of the average of the $E_0(\lambda)$. If $q \in \mathcal{C}_c^\infty((0, \infty))$ is a smooth function of compact support set $E_0(q)f = \int_0^\infty q(\lambda)E_0(\lambda)f d\lambda$. Then $E_0(q) : \mathcal{S}(\mathbb{R}^n) \longrightarrow \mathcal{S}(\mathbb{R}^n)$ and for any two functions q, $q' \in \mathcal{C}^\infty((0, \infty))$ it is always the case that $E_0(q) \circ E_0(q') = E_0(qq')$.

[9]   As a general convention I use the same notation for the operator and its Schwartz kernel. Of course this is a possible source of confusion and error; in particular one has to be careful as to which variables are regarded as parameters and what is the splitting into 'incoming' and 'outgoing' variables. Nevertheless I feel that this danger is outweighed by the consequent reduction in the number of symbols.

[10]   So one can reasonably say that $\Phi_0^*(z, \omega, \lambda) = \overline{\Phi^\dagger(z, \omega, \lambda)}$, where the $\dagger$ tells one to reverse the order of the variables and so get the transpose.

[11]   As follows from the properties of the Fourier transform, since any smooth function on the sphere $|\zeta| = \lambda > 0$ is the restriction of an element of $\mathcal{S}(\mathbb{R}^n)$.

Lemma 1.1 [12] *For $0 \neq \lambda \in \mathbb{R}$ the range of $\Phi_0(\lambda)$, acting on distributions on $\mathbb{S}^{n-1}$, is the null space of $\Delta - \lambda^2$ acting on the space, $\mathcal{S}'(\mathbb{R}^n)$, of tempered distributions on $\mathbb{R}^n$.*

1.3 Scattering matrix

Thus all the solutions of $(\Delta - \lambda^2)u = 0$ with u 'of polynomial growth' are superpositions of the elementary plane wave solutions $\Phi_0(z, \omega, \lambda) = e^{i\lambda z \cdot \omega}$ where $\omega \in \mathbb{S}^{n-1}$. The plane waves give a 'continuous'[13] parametrization' of the eigenspace; there is a related 'functional parametrization' of it which is also important.

If g in (1.9) is taken to be smooth then the principle of stationary phase [14] can be used to understand the behaviour of $\Phi_0(\lambda)g$ as $|z| \to \infty$. Writing $z = |z|\theta$, $\theta = z/|z| \in \mathbb{S}^{n-1}$ gives

$$(1.12) \qquad \Phi_0(\lambda)g(|z|\theta) = \int_{\mathbb{S}^{n-1}} e^{i|z|\lambda\theta \cdot \omega} g(\omega) d\omega.$$

The phase function $\theta \cdot \omega$ as a function of $\omega \in \mathbb{S}^{n-1}$ is stationary, i.e. has vanishing gradient, exactly at the two points $\omega = \pm\theta$. Since the Hessian at these points is non-degenerate[15] the stationary phase lemma gives a complete asymptotic expansion[16]

(1.13)

$$\Phi_0(\lambda)g(|z|\theta) \sim e^{i\lambda|z|} \left(\lambda|z|\right)^{-\frac{1}{2}(n-1)} e^{-\frac{1}{4}\pi(n-1)i} (2\pi)^{\frac{1}{2}(n-1)} \sum_{j \geq 0} |z|^{-j} h_j^+(\theta)$$

$$+ e^{-i\lambda|z|} \left(\lambda|z|\right)^{-\frac{1}{2}(n-1)} e^{\frac{1}{4}\pi(n-1)i} (2\pi)^{\frac{1}{2}(n-1)} \sum_{j \geq 0} |z|^{-j} h_j^-(\theta), \quad \lambda > 0,$$

---

[12] This is simple to prove using the structure theory of distributions. Namely if $(\Delta - \lambda^2)u = 0$ with $u \in \mathcal{S}'(\mathbb{R}^n)$, the dual space to $\mathcal{S}(\mathbb{R}^n)$, then the Fourier transform $\widehat{u}(\zeta)$ satisfies $(|\zeta|^2 - \lambda^2)\widehat{u}(\zeta) = 0$. If $\lambda \neq 0$ it follows that, written in terms of polar coordinates $z = r\theta$, $\widehat{u} = \delta(r - |\lambda|)g'(\theta)$ for some distribution on the sphere $g' \in C^{-\infty}(\mathbb{S}^{n-1})$. The inverse Fourier transform then shows that u is $\Phi_0(\lambda)g$ for $g = (2\pi)^{-n}\lambda^{n-1}g'$.

[13] Really this is a smooth parametrization. One view of scattering theory is that it describes the smoothness of the spectrum of appropriate operators.

[14] By a 'principle' here is meant an old theorem which has had many manifestations. For a precise statement of an appropriate version see [42], Section 7.7.

[15] That is, $\omega \cdot \theta$ is a Morse function on the sphere.

[16] This means that for any integer N the difference between the left side and the partial sum over $j \leq N$ on the right side is bounded, in $|z| \geq 1$, by $C|z|^{-N-1-\frac{1}{2}(n-1)}$ for some constant C. The power here is just the size of the first term dropped from the sums. In fact the same is true after any number of formal derivatives with respect to θ, or $r = |z|$, are taken (on both sides of course).

in which $h_0^\pm(\theta) = g(\pm\theta)$ and the $h_j^\pm$ for $j \geq 1$ are all given by polynomials in the Laplacian on the sphere applied to $g(\pm\theta)$.

Lemma 1.2 [17] *For each $\lambda > 0$ and each $h \in C^\infty(\mathbb{S}^{n-1})$ there is a unique solution to $(\Delta - \lambda^2)u = 0$ such that as $|z| \to \infty$* [18]

(1.15)
$$u(|z|\theta) = e^{i\lambda|z|}|z|^{-\frac{1}{2}(n-1)}h(\theta)$$
$$+ e^{-i\lambda|z|}|z|^{-\frac{1}{2}(n-1)}h'(\theta) + O\left(|z|^{-\frac{1}{2}(n+1)}\right)$$

where $h' \in C^\infty(\mathbb{S}^{n-1})$, and necessarily [19]

(1.16)
$$h'(\theta) = A_0 h(\theta) = i^{n-1}h(-\theta).$$

This parametrizes the generalized eigenspace with eigenvalue λ^2 by the distributions[20] on the sphere at infinity. Notice that $\pm\lambda$ give different parametrizations of the same space, one in terms of h and the other in terms of h'. The relationship between these two parametrizations is given by (1.16) and this operator, mapping $h(\theta)$ to $i^{(n-1)}h(-\theta)$, is the 'absolute scattering matrix' for Euclidean space. It is a unitary isomorphism of $C^\infty(\mathbb{S}^{n-1})$. [21]

There are various stronger forms of this lemma, as far as the uniqueness part is concerned. One particularly convenient one arises from the

[17] The existence part follows from (1.13). To prove the uniqueness it is only necessary to prove a variant of (1.13) for $\Phi_0(\lambda)g$ where $g \in C^{-\infty}(\mathbb{S}^{n-1})$ is a distribution. This can be done by using the same formula, (1.12), integrated against a test function in $C^\infty(\mathbb{S}^{n-1})$. The stationary phase expansion in the θ variable shows that

(1.14)
$$\Phi_0(\lambda)g(|z|\theta) = e^{i\lambda|z|}(\lambda|z|)^{-\frac{1}{2}(n-1)}e^{-\frac{1}{4}\pi(n-1)i}(2\pi)^{\frac{1}{2}(n-1)}g(\theta)$$
$$+ e^{-i\lambda|z|}(\lambda|z|)^{-\frac{1}{2}(n-1)}e^{\frac{1}{4}\pi(n-1)i}(2\pi)^{\frac{1}{2}(n-1)}g(-\theta) + u'$$

where $u' \in H^{-\infty}(\mathbb{R}^n)$ is in the union of all the standard Sobolev spaces; moreover g is determined by this expansion since neither the first two terms separately, not their sum, can be in $H^{-\infty}(\mathbb{R}^n)$ unless $g = 0$. Given two solutions of the form (1.15), the difference is a solution with $h = 0$. From Lemma 1.1 it follows that this difference is of the form $\Phi_0(\lambda)g$ for some g. The uniqueness of the expansion (1.14) then shows that $g = 0$. Some further comments on the uniqueness will be made in Lecture 2.

[18] The 'big Oh' notation here means that the difference of the left and right sides is bounded by $C|z|^{-\frac{1}{2}(n+1)}$ in $|z| \geq 1$ for some constant C.

[19] As defined here the operator A_0 is independent of λ. However, there is also a unique solution on the form (1.15) for $\lambda < 0$. If n is odd, the resulting operator mapping h to h' is the same. If n is even it is not, rather it is $-A_0$.

[20] I mean here that the map $h \mapsto u \in \mathcal{S}'(\mathbb{R}^n)$ extends by continuity to all $h \in C^{-\infty}(\mathbb{S}^{n-1})$ and then gives a parametrization of all the tempered generalized eigenfunctions.

[21] If n is odd it is an involution, i.e. $A_0 \circ A_0 = \mathrm{Id}$, whereas if n is even it is a fourth root of unity in the sense that $A_0^4 = \mathrm{Id}$. This sort of behaviour, depending on the parity of the dimension, can be seen much more strongly in 1.6.

observation that the function $|z|^{-\frac{1}{2}(n-1)}$ is locally square-integrable near 0 and that $|z|^{-\frac{1}{2}(n+1)}$ is square-integrable near $|z| = \infty$.[22] Thus (1.15) implies that

$$
\begin{aligned}
(1.17) \quad u(|z|\theta) =& e^{i\lambda|z|}|z|^{-\frac{1}{2}(n-1)}h(\theta) \\
&+ e^{-i\lambda|z|}|z|^{-\frac{1}{2}(n-1)}h'(\theta) + u', \ u' \in L^2(\mathbb{R}^n).
\end{aligned}
$$

Conversely, for a solution to $(\Delta - \lambda^2)u = 0$, this implies (1.15) and hence (1.13).

Notice from (1.13) that the map $\mathcal{C}^\infty(\mathbb{S}^{n-1}) \longmapsto \mathcal{S}'(\mathbb{R}^n)$ which gives the unique solution of the form (1.15) is

(1.18)
$$
u(z) = P_0(\lambda)h = \lambda^{\frac{1}{2}(n-1)} e^{\frac{1}{4}\pi(n-1)i}(2\pi)^{-\frac{1}{2}(n-1)}\Phi_0(\lambda)h, \ \lambda > 0.
$$

It would be reasonable to call the operator $P_0(\lambda)$ the 'Poisson operator' for the 'boundary problem' which seeks the solution to $(\Delta - \lambda^2)u = 0$ of the form (1.15) with h given.[23]

1.4 Resolvent family

I should pay at least lip service to the fundamental fact that the Laplacian is an essentially self-adjoint operator.[24] In particular the inverse of the operator $\Delta - \sigma$, for $\sigma \in \mathbb{C} \setminus \mathbb{R}$ is a bounded operator on $L^2(\mathbb{R}^n)$. This is certainly true and much more can be seen, namely that this operator can be obtained in terms of the Fourier transform:

$$
(1.19) \quad (\Delta - \sigma)^{-1}f(z) = (2\pi)^{-n} \int_{\mathbb{R}^n} e^{iz\cdot\zeta}(|\zeta|^2 - \sigma)^{-1}\hat{f}(\zeta)d\zeta
$$

whenever $\sigma \in \mathbb{C} \setminus [0, \infty)$. [25]

Since the spectrum [26] is confined to the positive real axis it is convenient to introduce $\lambda^2 = \sigma$ as a modified spectral parameter. There are two obvious normalizations of the choice of λ; I shall choose the 'physical

---

[22] That is, the function is square-integrable on the complement of any ball of positive radius around the origin.

[23] The mapping properties of an operator such as $P_0(\lambda)$ can be understood in terms of Besov spaces, see [43].

[24] If you want to know what this means see [98].

[25] Since then $|\zeta|^2 - \sigma$ has no zeroes for $\zeta \in \mathbb{R}^n$.

[26] The spectrum is the singular set of the resolvent family .

domain' to be the set[27]

(1.20) $$\mathcal{P} = \{\lambda \in \mathbb{C}; \operatorname{Im} \lambda < 0\}.$$

Then define

(1.21) $$R_0(\lambda) = (\Delta - \lambda^2)^{-1}, \ \lambda \in \mathcal{P}, \ \text{i.e. } \operatorname{Im} \lambda < 0.$$

I will usually refer to this, slightly incorrectly,[28] as 'the resolvent.' From (1.19) it follows that

(1.22) $$R_0(\lambda) : \mathcal{S}(\mathbb{R}^n) \longrightarrow \mathcal{S}(\mathbb{R}^n) \ \text{for } \operatorname{Im} \lambda < 0.$$

It is the unique operator with this property such that $(\Delta - \lambda^2) \circ R_0(\lambda) =$ Id on $\mathcal{S}(\mathbb{R}^n)$.

1.5 Limiting absorption principle

The resolvent of the Laplacian can be written as an integral operator:

(1.23)
$$R_0(\lambda) f(z) = \int_{\mathbb{R}^n} R_0(\lambda, z, z') f(z') dz',$$
$$R_0(\lambda, z, z') = (2\pi)^{-n} \int_{\mathbb{R}^n} e^{i(z-z')\cdot\zeta} \frac{d\zeta}{(|\zeta|^2 - \lambda^2)}.$$

The integral here is not absolutely convergent.[29] To avoid worrying about this[30] I shall consider instead the kth power of the resolvent, where for $k > \frac{1}{2}n$ the corresponding integral *is* absolutely convergent

(1.27) $$R_0^k(\lambda, z, z') = (2\pi)^{-n} \int_{\mathbb{R}^n} e^{i(z-z')\cdot\zeta} \frac{d\zeta}{(|\zeta|^2 - \lambda^2)^k}, \ \operatorname{Im} \lambda < 0.$$

---

[27] In the lectures themselves I used the opposite convention, that $\operatorname{Im} \lambda > 0$ in the physical domain, I regretted it then .... I hope that all the sign errors have been eliminated, but I am not too confident.

[28] In that the resolvent is $(\Delta - \sigma)^{-1}$.

[29] It is relatively straightforward to compute the form of these kernels 'explicitly'; the result (as with almost everything else) is simpler in the odd-dimensional case than the even-dimensional one. If $n = 1$ then

(1.24) $$R_0(\lambda, z, z') = \lambda^{-1} \exp(-i\lambda|z - z'|).$$

If $n \geq 3$ is odd then there is a polynomial, q_n, of degree $\frac{1}{2}(n-1)$ in one variable such that

(1.25) $$R_0(\lambda; z, z') = |z - z'|^{-n+2} q_n(\lambda|z - z'|) \exp(i\lambda|z - z'|).$$

For $n \geq 2$ even

(1.26) $$R_0(\lambda; z, z') = \frac{1}{4i} \left(\frac{\lambda}{2\pi|z - z'|}\right)^{\frac{1}{2}n-1} \mathrm{Ha}^{(1)}_{\frac{1}{2}n-1}(\lambda|z - z'|)$$

where $\mathrm{Ha}^{(1)}_j(z)$ is a Hankel function.

[30] Not that it is a serious problem.

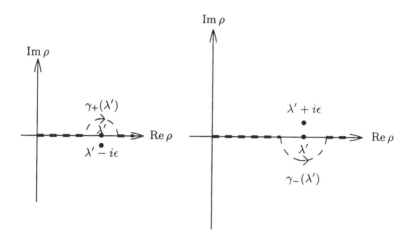

Fig. 1. The contours $\gamma_+(\lambda')$ and $\gamma_-(\lambda')$.

I am especially interested in what happens as $\operatorname{Im}\lambda \uparrow 0$ and the spectral parameter $\sigma = \lambda^2$ approaches $[0,\infty)$. Introducing polar coordinates in ζ, as in (1.7), but now writing $\zeta = \rho\omega$, gives

$$(1.28)\quad R_0^k(\lambda, z, z') = (2\pi)^{-n} \int_{\mathbb{S}^{n-1}} \int_0^\infty e^{i\rho(z-z')\cdot\omega} \frac{\rho^{n-1}d\rho}{(\rho^2 - \lambda^2)^k} d\omega.$$

The integrand is holomorphic in ρ away from $\rho = \pm\lambda$, where there is a pole. If $\lambda = \lambda' - i\epsilon$ with $\lambda' > 0$ and $\epsilon > 0$ and small, then Cauchy's theorem can be used to move the contour in (1.28) to $\gamma_+(\lambda')$ as in Figure 1:

(1.29)

$$R_0^k(\lambda, z, z') = (2\pi)^{-n} \int_{\mathbb{S}^{n-1}} \int_{\gamma_+(\lambda')} e^{i\rho(z-z')\cdot\omega} \frac{\rho^{n-1}d\rho}{(\rho^2 - \lambda^2)^k} d\omega, \quad \lambda = \lambda' - i\epsilon.$$

Now the limit as $\epsilon \downarrow 0$ in (1.29) is not singular, provided $\lambda' > 0$. If $\lambda = -\lambda' - i\epsilon$ where λ' is still positive then in place of (1.29)

(1.30)

$$R_0^k(\lambda, z, z') = (2\pi)^{-n} \int_{\mathbb{S}^{n-1}} \int_{\gamma_-(\lambda')} e^{i\rho(z-z')\cdot\omega} \frac{\rho^{n-1}d\rho}{(\rho^2 - \lambda^2)^k} d\omega,$$

$$\lambda = -\lambda' - i\epsilon,$$

with $\gamma_-(\lambda')$ the contour going 'the other way' around λ'. Again the limit as $\epsilon \downarrow 0$ can be taken. Since the two limiting points $\pm\lambda'$ correspond to the same point $\sigma = (\lambda')^2$ in the spectrum it is natural to consider the difference:

$$R_0^k(\lambda, z, z') - R_0^k(-\lambda, z, z') =$$

$$(1.31) \quad (2\pi)^{-n} \int_{\mathbb{S}^{n-1}} \int_{\gamma_+(\lambda)-\gamma_-(\lambda)} e^{i\rho(z-z')\cdot\omega} \frac{\rho^{n-1}d\rho}{(\rho^2 - \lambda^2)^k} d\omega, \quad \lambda > 0.$$

The difference of the two contours is homotopic to a clockwise circle of small radius around the single point λ, so Cauchy's theorem can be used to evaluate the integral as a residue. Using the identity $(\Delta - \lambda^2)^j R_0^k = R_0^{k-j}(\lambda)$ for $k > j$

$$R_0(\lambda; z, z') - R_0(-\lambda; z, z')$$

$$(1.32) \qquad = \frac{1}{2i}(2\pi)^{-(n-1)}\lambda^{n-2} \int_{\mathbb{S}^{n-1}} e^{i\lambda(z-z')\cdot\omega} d\omega, \quad \lambda > 0.$$

This is the limiting absorption principle or it can be called, perhaps more correctly, Stone's theorem. [31]

1.6 Analytic continuation

Formulæ (1.29) and (1.30) make sense for ϵ small compared to the radius of the circular part of the contour (and hence with respect to λ'), of either sign. This shows that $R_0(\lambda, z, z')$ can be continued analytically, as a function of λ, through the real axis, at least away from 0. Indeed if I define $M(\lambda)$, in terms of right side of (1.32):

$$(1.34) \quad M(\lambda, z, z') = \frac{1}{2i}(2\pi)^{-(n-1)} \int_{\mathbb{S}^{n-1}} e^{i\lambda(z-z')\cdot\omega} d\omega, \quad \lambda > 0,$$

then observe that $M(\lambda; z, z')$ extends to be an entire function of $\lambda \in \mathbb{C}$. I shall denote, temporarily, by $\widetilde{R}_0(\lambda)$ the function [32] defined by analytic continuation of $R_0(\lambda)$ across $(0, \infty)$. Thus $\widetilde{R}_0(\lambda) = R_0(\lambda)$ in $\text{Im}\,\lambda < 0$,

---

[31] Which can be stated briefly in the present context as the assertion that the spectral resolution can be obtained from the difference of the limits, from above and below, of the resolvent family on the spectrum. Notice that by inserting the Fourier transform in (1.7) it follows that

$$(1.33) \qquad E_0(\lambda) = \frac{\lambda i}{\pi} (R_0(\lambda) - R_0(-\lambda)), \quad \lambda \in (0, \infty).$$

---

[32] Really to be thought of as an operator.

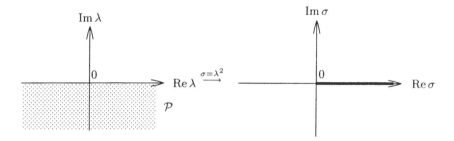

Fig. 2. Analytic continuation of the resolvent for n odd.

but $\widetilde{R}_0(\lambda)$ is also defined near the positive real axis. From (1.32) it follows that, for λ near the positive real axis with $\mathrm{Im}\,\lambda > 0$,

$$(1.35) \qquad \widetilde{R}_0(\lambda) = R_0(-\lambda) + \lambda^{n-2} M(\lambda).$$

Thus in fact $\widetilde{R}_0(\lambda)$ extends to be holomorphic for all $\mathrm{Im}\,\lambda > 0$ as well as near $(0, \infty)$ and in $\mathrm{Im}\,\lambda < 0$, i.e. $\widetilde{R}_0(\lambda)$ is holomorphic in $\mathbb{C} \setminus (-\infty, 0]$.

Using the antipodal map in the sphere it follows from (1.34) that

$$(1.36) \qquad M(-\lambda) = M(\lambda).$$

Applying (1.35) twice gives

$$\lim_{\epsilon \downarrow 0} \widetilde{R}_0(-\lambda' + i\epsilon) - \lim_{\epsilon \uparrow 0} \widetilde{R}_0(-\lambda' + i\epsilon)$$

$$(1.37)$$

$$= \begin{cases} 0 & n \text{ odd} \\ 2(\lambda')^{n-2} M(\lambda') & n \text{ even} \end{cases} \quad \lambda' > 0,$$

This shows the basic difference between the odd- and even-dimensional cases. For odd $n \geq 3$, the resolvent kernel is locally integrable in z, z' and entire[33] as a function of λ; (see Figure 2) for $n = 1$ it is meromorphic in λ with a simple pole at 0. In the even-dimensional case a similar result is valid, except that the kernel only extends to be entire on the logarithmic covering of the complex plane, Λ, i.e. as a function of the variable $\log \lambda$ (see Figure 3).[34] Thus if $R_0^\flat(\tau) = R_0(\lambda)$ the 'physical domain' for $R_0^\flat(\tau)$ can be taken as $\{\tau \in \mathbb{R} \times i(-\pi, 0) \subset \mathbb{C}\}$ and then $R_0^\flat(\tau)$ extends to

[33] It follows from (1.23) that there is neither an essential singularity, nor a pole, at $\lambda = 0$.

[34] In the even dimensional case the behaviour as $\lambda \to 0$ can also be analyzed; in fact

$$(1.38) \qquad R_0(\lambda) = R_0'(\lambda) + M(\lambda)\lambda^{n-2} \log \lambda$$

where $R_0'(\lambda)$ is entire.

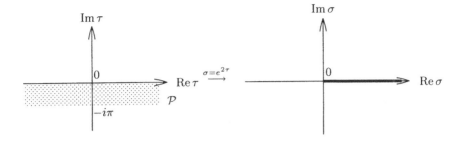

Fig. 3. Analytic continuation of the resolvent for n even.

be entire on the τ-plane and has the special property that under the transformation $\tau \mapsto \tau + \pi i$,[35] which corresponds to the shift from one preimage of the point $\lambda = e^{\tau}$ to another, it transforms by

$$(1.39) \qquad R_0^{\flat}(\tau + \pi i) = R_0^{\flat}(\tau) + e^{(n-2)\tau} M(e^{\tau})$$

where $M(\lambda)$ is the entire function of λ given by (1.34).[36] In either case I shall denote the analytic continuation again by $R_0(\lambda)$, even though in the even-dimensional case it is a function on Λ.

1.7 Asymptotic expansion

The expansion, (1.15), for elements of the null space of $\Delta - \lambda^2$ can be extended to elements of the 'near null space.' More precisely

Proposition 1.1 *If* $\lambda \in \mathcal{P} \cup (\mathbb{R} \setminus \{0\})$[37] *then for each* $f \in \mathcal{C}_c^{\infty}(\mathbb{R}^n)$[38]

$$(1.41)$$

$$(R_0(\lambda)) f(|z|\theta) \sim e^{-i\lambda|z|} |z|^{-\frac{1}{2}(n-1)} \sum_{j=0}^{\infty} |z|^{-j} h_j(\theta) \ as \ |z| \to \infty,$$

---

[35] This is the 'deck-transformation' for the covering of the λ-plane by the τ-plane.

[36] Note that $M(\lambda)$ can be expressed, for real $\lambda > 0$, as

$$(1.40) \qquad M(\lambda) = \frac{1}{2i} (2\pi)^{-(n-1)} \Phi_0(\lambda) \Phi_0^*(\lambda).$$

Indeed this follows directly from (1.34) and (1.9), or (1.33).

[37] Thus, $\mathrm{Im}\,\lambda \leq 0$ and $\lambda \neq 0$.

[38] For real λ this result remains true for $f \in \mathcal{S}(\mathbb{R}^n)$, although the proof is a little more involved.

with $h_j \in C^\infty(\mathbb{S}^{n-1})$ and where[39]

(1.42)
$$h_0(\theta) = \frac{1}{2i\lambda} P_0^\dagger(\lambda) f = \frac{1}{2i}(2\pi)^{-\frac{1}{2}(n+1)} \lambda^{\frac{1}{2}(n-3)} e^{\frac{1}{4}\pi(n-1)i} \widehat{f}(-\lambda\theta), \quad \lambda > 0.$$

This result can be proved using methods similar to those discussed in Section 1.5.[40] If $\lambda > 0$ then only the first term in (1.41) is not square-integrable near infinity. The solution $u = R_0(\lambda)f$ to $(\Delta - \lambda^2)u = f$ is then distinguished by the *Sommerfeld radiation condition:*

(1.46)
$$\left(\frac{\partial}{\partial r} + i\lambda\right) u(r\theta) \in L^2(\mathbb{R}^n).$$

---

[39] Here $P_0^\dagger(\lambda)$ is the transpose of the operator $P_0(\lambda)$ in (1.18), so has the same kernel but with the variables reversed in order.

[40] I shall only discuss the proof for real λ. Consider $R_0(\lambda)$ which I have defined as the limit of the resolvent from the physical region. Choose a cut-off function $\phi \in C_c^\infty(\mathbb{R})$ which is 1 in an open neighbourhood of 1 and vanishes identically near 0. Then, for $\mathrm{Im}\,\lambda < 0$,

$$R_0(\lambda)f(z) = u_1(z) + u_2(z),$$

(1.43)
$$\widehat{u_1}(\zeta) = \phi\left(\frac{|\zeta|}{|\lambda|}\right)(|\zeta|^2 - \lambda^2)^{-1}\widehat{f}(\zeta).$$

Here, $u_2 \in \mathcal{S}(\mathbb{R}^n)$ so it remains to analyze the behaviour of u_1. Now fix $\theta = z/|z|$ and consider another cut-off function on the sphere, $\psi_0 \in C^\infty(\mathbb{S}^{n-1})$ such that $\psi_0(\omega)$ is identically equal to 1 in a neighbourhood of the equator $\omega \cdot \theta = 0$ and with $\psi_0(\omega)$ vanishing identically in neighbourhoods of both $\omega = \theta$ and $\omega = -\theta$. Let $1 = \psi_0 + \psi_+ + \psi_-$ be the resulting partition of unity with $\psi_\pm$ supported in $\pm\theta \cdot \omega \geq 0$. Writing u_1 as the inverse Fourier transform of its Fourier transform, introducing polar coordinates $\zeta = \rho\omega$ and inserting this partition of unity gives

(1.44)
$$u_1(z) = v_0(z) + v_+(z) + v_-(z) \text{ where}$$

$$v_t = (2\pi)^{-n} \int_0^\infty \int_{\mathbb{S}^{n-1}} e^{i|z|\rho\omega\cdot\theta}\psi_t(\omega)(\rho^2 - \lambda^2)^{-1}\rho^{n-1}\phi\left(\frac{\rho}{|\lambda|}\right)\widehat{f}(\rho\omega)d\omega d\rho.$$

Of the three terms in (1.44) the only non-trivial one, asymptotically, is v_-, since $v_0, v_+ \in \mathcal{S}(\mathbb{R}^n)$ uniformly for $\mathrm{Im}\,\lambda \leq 0$ (but near a fixed non-zero real value λ'.) For v_0 this follows by integration by parts in ω, using the fact that $d_\omega(\omega \cdot \theta) \neq 0$ on the support of ψ_0. For v_+ the ρ-integral can be deformed into the complex, as $\gamma_+(\lambda')$, using the fact that the integrand is analytic in ρ in a neighbourhood of the deformed part. Similarly the ρ integral for v_- can be deformed to a contour integral over $\gamma_-(\lambda')$ except that the pole at $\rho = \lambda$ is encountered during the deformation. Thus

(1.45)
$$R_0(\lambda)f(|z|\theta) = \frac{1}{2}(2\pi)^{-n}\lambda^{n-2}\int_{\mathbb{S}^{n-1}} e^{i\lambda|z|\omega\cdot\theta}\psi_-(\omega)\widehat{f}(\lambda\omega)d\omega + w, \quad w \in \mathcal{S}(\mathbb{R}^n).$$

The asymptotic expansion then follows from the principle of stationary phase, much as in (1.13).

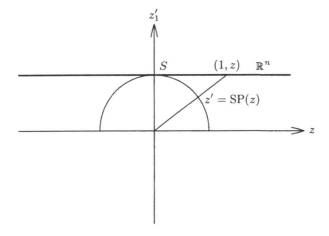

Fig. 4. Stereographic, or radial, compactification of $\mathbb{R}^n$.

1.8 Radial compactification

One point of view that I would like to emphasize from the beginning of these lectures is that non-compact spaces, such as $\mathbb{R}^n$, should generally be compactified. The idea here is simply that I do not want to think of asymptotic expansions such as (1.13) as some new phenomenon taking place 'at infinity.' Rather this is just a form of Taylor's theorem at the boundary (which is 'infinity'). To see this just carry out the stereographic, or perhaps more correctly radial, compactification of Euclidean space, $\mathbb{R}^n$, to a ball, or better yet a half-sphere as in Figure 4

$$(1.47) \qquad \mathbb{S}^n_+ = \{z' \in \mathbb{R}^{n+1}; |z'| \leq 1,\ z'_1 \geq 0\}.$$

Stereographic projection is the identification of $\mathbb{R}^n$ with the interior of the half-sphere:

$$(1.48)$$

$$\mathrm{SP} : \mathbb{R}^n \ni z \longmapsto z' = \left(\frac{1}{(1+|z|^2)^{\frac{1}{2}}}, \frac{z}{(1+|z|^2)^{\frac{1}{2}}} \right) \in \mathbb{S}^n_+ \subset \mathbb{R}^{n+1}.$$

I shall consistently denote by x a defining function[41] for the boundary of a manifold with boundary. In this case $x = z'_1 = (1+|z|^2)^{-\frac{1}{2}}$ is a defining function for the boundary of $\mathbb{S}^n_+$. In $|z| > 1$, $(1+|z|^2)^{\frac{1}{2}} = |z|^{-1}(1+|z|^{-2})^{\frac{1}{2}}$

---

[41] A defining function for a hypersurface H in a manifold M is a real-valued function $\rho \in \mathcal{C}^\infty(M)$ which vanishes precisely on H, so $H = \{p \in M; \rho(p) = 0\}$ and has $d\rho(p) \neq 0$ at all points of H. For the boundary of a manifold with boundary I shall assume that x is normalized to be positive in the interior of M.

and it follows that $|z|^{-1}$ is a boundary defining function for $\mathbb{S}^n_+$, except for the minor problem that it blows up at the interior point corresponding to the origin in $\mathbb{R}^n$. This means that (1.13) can be rewritten in the form[42]

$$\Phi_0(\lambda)g = \mathrm{SP}^* f \text{ where}$$

(1.49)
$$f = e^{i\lambda/x} x^{\frac{1}{2}(n-1)} h_+ + e^{-i\lambda/x} x^{\frac{1}{2}(n-1)} h_-$$

$$\text{with } h_\pm \in \mathcal{C}^\infty(\mathbb{S}^n_+).$$

The asymptotic expansion (1.13) follows from the Taylor expansion of the functions $h_\pm$[43] at the boundary of $\mathbb{S}^n_+$.[44] The Sommerfeld radiation condition can be written

(1.50)
$$(x^2 \frac{\partial}{\partial x} - i\lambda)u \in L^2_{\mathrm{sc}}(\mathbb{S}^n_+)$$

where $L^2_{\mathrm{sc}}(\mathbb{S}^n_+)$ is the space of square-integrable functions for the metric volume form, $L^2(\mathbb{R}^n) = \mathrm{SP}^* L^2_{\mathrm{sc}}(\mathbb{S}^n_+)$.

I will note here, for later reference, the form of the Euclidean metric as a metric on the interior of $\mathbb{S}^n_+$. Introducing polar coordinates on $\mathbb{R}^n$, $\theta = z/|z|$, $R = |z|$ the metric becomes

(1.51)
$$|dz|^2 = dR^2 + R^2|d\theta|^2$$

where $|d\theta|^2$ denotes the usual metric on the sphere $\mathbb{S}^{n-1}$. If $x = |z|^{-1}$ is the boundary defining function discussed above then the metric can be written, near the boundary, in the form

(1.52)
$$|dz|^2 = \frac{dx^2}{x^4} + \frac{|d\theta|^2}{x^2}.$$

Generalizations of this type of metric, and the associated scattering theory, to an arbitrary compact manifold with boundary in place of $\mathbb{S}^n_+$ will be discussed in Lecture 6.

---

[42] For f a function on $\mathbb{S}^n_+$, the pull-back to $\mathbb{R}^n$ is $\mathrm{SP}^* f = f \circ \mathrm{SP}$.

[43] Notice that $\mathcal{C}^\infty(\mathbb{S}^n_+)$ is the space of functions on $\mathbb{S}^n_+$ which are continuous up to the boundary with all their derivatives. Thus demanding $h \in \mathcal{C}^\infty(\mathbb{S}^n_+)$ is the same as saying that $h = \widetilde{h} \restriction \mathbb{S}^n_+$ for some function $\widetilde{h} \in \mathcal{C}^\infty(\mathbb{S}^n)$.

[44] Conversely (1.49) can be deduced from (1.13), together with similar estimates on formal derivatives of the expansion.

2

Potential scattering on $\mathbb{R}^n$

The simplest perturbations of the flat Laplacian on Euclidean space are given by potentials. I will spend this second lecture showing the degree to which the results I described last time, for the flat case, extend when the Laplacian is perturbed in this way. One reason I wish to concentrate on potential perturbations is their simplicity, which means that I can even outline the methods of proof. Much of what I will say carries over to other perturbations and I shall say a little more about this later.

So consider the operator $\Delta + V$ where $V \in \mathcal{C}_c^\infty(\mathbb{R}^n)$. Here V acts by multiplication and is the 'potential.' [1] I shall limit myself to this simple case, where V is both smooth and has compact support, even though most of the results I describe have generalizations involving less regularity or less stringent support properties (e.g. replaced by growth conditions at infinity). In fact much energy has gone into refining these results for low regularity potentials with weak decay conditions at infinity.

2.1 The resolvent of $\Delta + V$

The true spectral theory of $\Delta + V$ is very simple. Namely the new operator is almost unitarily equivalent to the free one. As in the free case, rather than discuss the technicalities of the fact that $\Delta + V$ is an unbounded self-adjoint[2] operator on $L^2(\mathbb{R}^n)$, I shall simply discuss the resolvent family.

---

[1] For the most part I shall assume that V is real-valued even though this is not always necessary.

[2] For real V. The notes of Simon's lectures [17] contain a good treatment of conditions on a potential (much more general than smooth with compact support) guaranteeing that $\Delta + V$ is self-adjoint.

Proposition 2.1 *For each* $\lambda \in \mathbb{C}$, $\text{Im}\,\lambda \ll 0$,[3] *there is a uniquely defined operator* $R_V(\lambda) : S(\mathbb{R}^n) \longrightarrow S(\mathbb{R}^n)$ *such that*

$$(2.1) \qquad\qquad (\varDelta + V - \lambda^2) \circ R_V(\lambda) = \text{Id}.$$

Indeed from the construction, via analytic Fredholm theory, that I will outline much more can be said about the family $R_V(\lambda)$. The radial compactification of $\mathbb{R}^n$ to a ball, or half-sphere, reduces $S(\mathbb{R}^n)$ to $\dot{C}^\infty(\mathbb{S}_+^n)$, the space of smooth functions on $\mathbb{S}_+^n$ vanishing with all derivatives at the boundary, i.e. $S(\mathbb{R}^n) = \text{SP}^* \dot{C}^\infty(\mathbb{S}_+^n)$. Thus, for $\text{Im}\,\lambda < 0$, $R_V(\lambda) : \dot{C}^\infty(\mathbb{S}_+^n) \longrightarrow \dot{C}^\infty(\mathbb{S}_+^n)$.[4]

Proposition 2.2 *For* $f \in C_c^\infty(\mathbb{R}^n) \subset \dot{C}^\infty(\mathbb{S}_+^n)$, $n \geq 2$,

$$(2.2) \qquad \begin{aligned} &R_V(\lambda)f = R_0(\lambda)G_V(\lambda)f \text{ with } G_V(\lambda) \text{ extending to be} \\ &\textit{meromorphic in } \lambda \text{ as a map } G_V(\lambda) : C_c^\infty(\mathbb{R}^n) \longrightarrow C_c^\infty(\mathbb{R}^n). \end{aligned}$$

Here $\lambda \in \mathbb{C}$ *if* n *is odd and* $\lambda \in \Lambda$ *if* n *is even.*

In fact even more can be said, namely that $G_V(\lambda)$ has finite rank residues at each pole and that these residues are smoothing operators, having kernels in $C^\infty(\mathbb{R}^n \times \mathbb{R}^n)$.

Proof The construction of $R_V(\lambda)$ proceeds via perturbation theory (which is why it is so easy) and then the extra properties follow from analytic Fredholm theory. Observe that, for $n > 1$, the corresponding operator $R_0(\lambda)$ has these properties and no poles at all.

Starting from the desired identity (2.1), and the corresponding free identity, it follows that

$$(2.3)$$
$$R_0(\lambda) = R_0(\lambda) \circ (\varDelta + V - \lambda^2) \circ R_V(\lambda) = R_V(\lambda) + R_0(\lambda) \circ V \circ R_V(\lambda).$$

This can be written $(\text{Id} + R_0(\lambda) \circ V) \circ R_V(\lambda) = R_0(\lambda)$. Without worrying for the moment about whether it makes sense, it is only necessary to invert the operator $\text{Id} + R_0(\lambda) \circ V$ and then[5]

$$(2.4) \qquad\qquad R_V(\lambda) = (\text{Id} + R_0(\lambda) \circ V)^{-1} R_0(\lambda).$$

---

[3] That is, $\text{Im}\,\lambda < c(V)$ for some constant $c(V)$ depending on V. There can be only a finite number of poles of $R_V(\lambda)$ in $\text{Im}\,\lambda < 0$. If V is real they must lie on the negative imaginary axis.

[4] I will generally identify an operator on $S(\mathbb{R}^n)$ with the operator on $\dot{C}^\infty(\mathbb{S}_+^n)$ to which it is conjugated by SP^*.

[5] This is often called the Lipmann-Schwinger equation, as are several other closely related equations.

Now, I need to do a little functional analysis to see this. First consider the Hilbert space $e^{T|z|}L^2(\mathbb{R}^n)$, meaning the space of functions of the form $e^{T|z|}f(z)$ where f is square-integrable. Let $L^2_c(B(R))$ be the space of square-integrable functions on $\mathbb{R}^n$ with support in the ball of radius $B(R) = \{|z| \le R\}$. In the region of $\mathbb{C}$ or Λ, depending on the parity of the dimension, where $|\operatorname{Im}\lambda| < T$, $R_0(\lambda)$ defines a family of compact operators [6]

$$(2.5) \qquad R_0(\lambda) : L^2_c(B(R)) \longrightarrow e^{T|z|}L^2(\mathbb{R}^n)$$

depending holomorphically on λ. Furthermore the norm of this operator tends to zero as $\lambda \to -i\infty$ in the original 'physical' half-plane, $\mathcal{P}$. Since V has compact support, if R is taken to be large enough then $V :$ $e^{T|z|}L^2(\mathbb{R}^n) \longrightarrow L^2_c(B(R))$. Now 'analytic Fredholm theory' [7] shows that the inverse family

$$(2.6) \qquad (\operatorname{Id} + R_0(\lambda) \circ V)^{-1} : e^{T|z|}L^2(\mathbb{R}^n) \longrightarrow e^{T|z|}L^2(\mathbb{R}^n)$$

is meromorphic with all residues of its inverse being operators of finite rank. In particular this inverse exists for all λ outside a discrete set. The uniqueness of the inverse in the physical region shows that enlarging T gives an extension of the same family. Finally this shows the existence of $R_V(\lambda)$, given by (2.4).[8]

I still need to check the stated properties of $R_V(\lambda)$, in particular (2.2). Instead of (2.3) the similar identity, arising from the fact that $R_V(\lambda)$ is expected to be a two-sided inverse, can be used:

$$(2.7)$$
$$R_0(\lambda) = R_V(\lambda) \circ (\Delta + V - \lambda^2) \circ R_0(\lambda) = R_V(\lambda) + R_V(\lambda) \circ V \circ R_0(\lambda).$$

Now the operator

$$(2.8) \qquad \operatorname{Id} + V \circ R_0(\lambda) : L^2_c(B(R)) \longrightarrow L^2_c(B(R))$$

---

[6] The exponential bound follows from the estimates on $R_0(\lambda)$ discussed in Lecture 1. The compactness is a form of the Ascoli-Arzela theorem, for the embedding of Sobolev spaces.

[7] The fact that a compact operator, such as $R_0 \circ V$, is norm-approximable by finite rank operators shows the invertibility of $\operatorname{Id} + R_0(\lambda) \circ V$ is, locally in λ, equivalent to the invertibility of a matrix and hence to the invertibility of a function (the determinant of the matrix.) Thus if the family is invertible at one point, as it is in this case, it is invertible outside a discrete set at which the inverse family has poles.

[8] The composition here makes sense since $(\operatorname{Id} + R_0(\lambda) \circ V)^{-1} : e^{T|z|}L^2(\mathbb{R}^n) \longrightarrow e^{T|z|}L^2(\mathbb{R}^n)$ for any T large enough compared with $|\lambda|$.

is entire with a meromorphic inverse which is just $G_V(\lambda)$. It is also the case that if $f \in \dot{C}^\infty(B(R))$ then[9]

$$(2.9) \qquad G_V(\lambda)f = (\mathrm{Id} + V \circ R_0(\lambda))^{-1} f \in \dot{C}^\infty(B(R))$$

depends meromorphically on λ. This leads to (2.2). That this construction gives the same operator $R_V(\lambda)$ follows from the invertibility near $-i\infty$.

2.2 Poles of the resolvent

The poles of the analytic continuation of the resolvent are in many ways similar to the eigenvalues of the Laplacian on a compact manifold with boundary, except that they are not real! They will be discussed at greater length in Lecture 4 but for the moment I simply note that they are associated to generalized eigenfunctions. Indeed it follows from (2.2) that, for n odd, if the resolvent R_V has a pole at λ then there is an eigenfunction u of the form

$$(2.10)$$
$$u = \exp(-i\lambda/x)x^{\frac{1}{2}(n-1)}w, \ \ w \in C^\infty(\mathbb{S}^n_+), \ \ (\varDelta + V - \lambda^2)u = 0.$$

In the even-dimensional case the same is true in the physical region, but the generalized eigenfunctions corresponding to poles of the resolvent are not quite so simple. However, they can be characterized in a uniform way:

Lemma 2.1 [10] *For $n \geq 2$, λ is a pole of $R_V(\lambda)$ if and only if there is a non-trivial solution, $u \in C^\infty(\mathbb{R}^n)$, to $(\varDelta + V - \lambda^2)u = 0$ such that $u = -R_0(\lambda)Vu$.*

If $\mathrm{Im}\,\lambda < 0$ then a function of the form (2.10) is square-integrable. If $\mathrm{Im}\,\lambda > 0$ it is not, nor is it for $0 \neq \lambda \in \mathbb{R}$, unless the coefficient, w,

---

[9] This is 'elliptic regularity.' The operator $R_0(\lambda)$ maps $H^p_c(\mathbb{R}^n)$ into $H^{p+2}_{\mathrm{loc}}(\mathbb{R}^n)$ for any λ and p, where $H^p_c(\mathbb{R}^n)$ and $H^p_{\mathrm{loc}}(\mathbb{R}^n)$ are, respectively, the spaces of functions with compact support in, and locally in, the Sobolev space $H^p(\mathbb{R}^n)$.

[10] This characterization follows directly from (2.2). Namely, if $R_V(\lambda)$ has a pole, then so must $G_V(\lambda)f$ for some $f \in C^\infty_c(\mathbb{R}^n)$. Since $(\mathrm{Id} + V \circ R_0(\lambda))G_V(\lambda) = \mathrm{Id}$ the residue, $u' \in C^\infty_c(\mathbb{R}^n)$, of $G_V(\lambda)f$ must satisfy $u' + VR_0(\lambda)u' = 0$. Then $u = R_0(\lambda)u'$ satisfies $u = -R_0(\lambda)VR_0(\lambda)u' = -R_0(\lambda)Vu$. Conversely if there is such a function $u \in C^\infty(\mathbb{R}^n)$, then $g = Vu \in C^\infty_c(\mathbb{R}^n)$ satisfies $g = -VR_0(\lambda)g$ which means that $\mathrm{Id} + V \circ R_0(\lambda)$ has null space and therefore cannot be invertible, so $G_V(\lambda)$, and hence $R_V(\lambda)$, must have a pole.

vanishes at the boundary. The real axis in λ is of particular interest since this gives rise to the spectrum of $\Delta + V$.

I shall summarize the 'algebraic' information in the poles in the following terms.

Definition 2.1 *Let $D(V) \subset \mathbb{C} \times \mathbb{N}$, for n odd, or $D(V) \subset \Lambda \times \mathbb{N}$ for n even, be the divisor defined by $R(\lambda)$. Thus $(\lambda', k) \in D(V)$ if $R(\lambda)$ has a pole at $\lambda = \lambda'$ of rank[11] k.*

2.3 Boundary pairing

As a continuation to Lemma 2.1 let me note the 'absence of embedded eigenvalues:'

Proposition 2.3 *For any n, and $V \in \mathcal{C}_c^\infty(\mathbb{R}^n)$ real-valued, there are no non-trivial solutions to (2.10) with $0 \neq \lambda \in \mathbb{R}$.*

Note that, for either parity of n, when $0 \neq \lambda \in \mathbb{R}$ the residue of any pole at λ must satisfy (2.10). There are two parts to the proof of Proposition 2.3. The first part is to show that if u satisfies (2.10) then it is actually in $\dot{\mathcal{C}}^\infty(\mathbb{S}_+^n)$, i.e. $\mathcal{S}(\mathbb{R}^n)$. This argument in turn consists of two steps. First it will be seen from a general 'boundary pairing,' which arises as a form of Green's formula, that $w = 0$ on $\mathbb{S}^{n-1} = \partial \mathbb{S}_+^n$. Then an inductive argument will be used to show that the whole Taylor series vanishes, so $u \in \dot{\mathcal{C}}^\infty(\mathbb{S}_+^n)$.

To define the boundary pairing, consider a more general 'formal solution' than in (2.10). Namely suppose that

(2.11)
$$u = u_+ + u_-, \quad u_\pm = \exp(\pm i\lambda/x)x^{\frac{1}{2}(n-1)}w_\pm, \quad w_\pm \in \mathcal{C}^\infty(\mathbb{S}_+^n) \text{ and}$$
$$(\Delta + V - \lambda^2)u = f \in \dot{\mathcal{C}}^\infty(\mathbb{S}_+^n).$$

---

[11] This is the algebraic multiplicity of the pole. In this context it can be defined as the dimension of the subspace of $\mathcal{C}^\infty(\mathbb{R}^n)$ which is the sum of the ranges of all the singular terms in the Laurent series for $R_V(\lambda)$ at $\lambda = \lambda'$. For $\lambda \neq 0$ this is the same as the range of the residue, i.e. the least singular term. All the elements of this space are annihilated by $(\Delta - (\lambda')^2)^k$. See also the paper of Gohberg and Sigal [28].

Potential scattering on $\mathbb{R}^n$

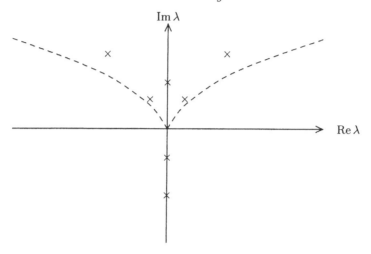

Fig. 5. Poles of the analytic continuation of $R_V(\lambda)$ (n odd).

Lemma 2.2 [12] *Suppose* $u^{(i)}$ *for* $i = 1, 2$ *are as in (2.11),* $V \in C_c^\infty(\mathbb{R}^n)$ *is real-valued and* $0 \neq \lambda \in \mathbb{R}$ *then*

(2.14)
$$-2i\lambda \int_{\mathbb{S}^{n-1}} \left(v_+^{(1)}\overline{v_+^{(2)}} - v_-^{(1)}\overline{v_-^{(2)}} \right) dz = \int_{\mathbb{R}^n} \left(f^{(1)}\overline{u^{(2)}} - u^{(1)}\overline{f^{(2)}} \right) dz$$

$$\text{where } v_\pm^{(i)} = w_\pm^{(i)} \restriction \partial \mathbb{S}_+^n.$$

Applying (2.14) directly to u as in (2.10) it follows that the leading coefficient $v = v_+ = w \restriction \partial \mathbb{S}_+^n$ vanishes identically.

---

[12] To prove this, choose a 'cut-off' function $\rho \in C_c^\infty(\mathbb{R})$ which has $\rho(x) = 1$ for $|x| < 1$ and support in $|x| \leq 2$. Now consider the integral

(2.12)
$$I_\epsilon = \int_{\mathbb{R}^n} \left(f^{(1)}\overline{u^{(2)}} - u^{(1)}\overline{f^{(2)}} \right) \rho(\epsilon|z|) dz.$$

Clearly, as $\epsilon \downarrow 0$ this converges to the right side of (2.14). Inserting $f^{(i)} = (\Delta + V - \lambda^2)u^{(i)}$ the terms involving V and λ^2 cancel. Integration by parts and use of the given form (2.11) shows that I_ϵ converges to the left side of (2.14). If V is not real then an alternative version of (2.14) is still available:

(2.13)
$$-2i\lambda \int_{\mathbb{S}^{n-1}} \left(v_+^{(1)} v_-^{(2)} - v_-^{(1)} v_+^{(2)} \right) = \int_{\mathbb{R}^n} \left(f^{(1)} u^{(2)} - u^{(1)} f^{(2)} \right) dz.$$

2.4 Formal solutions

Consider the structure of a formal solution as in (2.11). Since V has compact support this has nothing to do with V at all, i.e. $u_\pm$ are just formal solutions of the free Laplacian in the sense that $\Delta u_\pm = f_\pm \in \mathcal{S}(\mathbb{R}^n)$.

Lemma 2.3 [13] *For each $h \in \mathcal{C}^\infty(\mathbb{S}^{n-1})$ and $0 \neq \lambda \in \mathbb{R}$ there is an element, $u \in \mathcal{C}^\infty(\mathbb{R}^n)$, of the formal null space of $\Delta - \lambda^2$, i.e. $(\Delta - \lambda^2)u \in \mathcal{S}(\mathbb{R}^n)$, having an asymptotic expansion*

(2.15)
$$u \sim \exp(i\lambda|z|)|z|^{-\frac{1}{2}(n-1)} \sum_{j \geq 0} |z|^{-j} h_j(\theta), \ z = |z|\theta \text{ with } h_0 = h.$$

Moreover the difference of any two elements of the formal null space satisfying (2.15) is in $\mathcal{S}(\mathbb{R}^n)$.

2.5 Unique continuation

The remainder of the argument needed to prove Proposition 2.3 is a unique continuation theorem:

Theorem 2.1 [14] *If $0 \neq \lambda \in \mathbb{R}$ then any function $u \in \dot{\mathcal{C}}^\infty(\mathbb{S}^n_+)$ satisfying $(\Delta + V - \lambda^2)u = 0$ vanishes identically.*

---

[13] To see the existence of u consider (1.12) and (1.14). This certainly gives a formal solution (indeed a solution) of the form (2.11) with leading coefficient in u_+ being h. Using Borel's lemma (see [40], Theorem 1.2.6) at the boundary of $\mathbb{S}^n_+$, the coefficients $|z|^{-j}h_j^+ = x^j h_j^+$ can be summed, uniquely modulo $\mathcal{S}(\mathbb{R}^n) = \dot{\mathcal{C}}^\infty(\mathbb{S}^n_+)$, to give a solution of the form (2.15). The uniqueness follows by noting that if u has an expansion as in (2.15) with leading term, h, identically zero then u is square-integrable. Taking the Fourier transform of the equation $(\Delta - \lambda^2)u = f$ it follows that $(|\zeta|^2 - \lambda^2)^{-1}\hat{f}$ must be square-integrable and hence $\hat{f}$ must vanish on $\{|\zeta|^2 = \lambda^2\}$. This shows that $u \in \mathcal{S}(\mathbb{R}^n)$.

[14] This result is valid for complex-valued $V \in \mathcal{C}^\infty_c(\mathbb{R}^n)$. To prove it consider an eigenfunction $u \in \mathcal{S}(\mathbb{R}^n)$, which therefore satisfies $(\Delta - \lambda^2)u = -Vu = f \in \mathcal{C}^\infty_c(\mathbb{R}^n)$. The first step is to show that this implies that $u \in \mathcal{C}^\infty_c(\mathbb{R}^n)$, i.e. u vanishes near infinity. There are several ways to see this, one is to use the expansion of u in spherical harmonics. Each of the coefficients in this expansion is a function of $r = |z|$ which is rapidly decreasing and satisfies an ordinary differential equation. It can be seen that the equation is a form of Bessel's equation which has no non-trivial rapidly decreasing solution. Another proof can be based on the characterization of the range of the Radon transform and is briefly discussed in Footnote 3.5.
Once such a putative eigenfunction corresponding to an embedded eigenvalue is known to have compact support it can be shown to vanish identically using the unique continuation of solutions to second order elliptic equations.

This completes the proof of Proposition 2.3. Thus, if V is real-valued[15] then $R_V(\lambda)$ is well defined for all real, non-zero, λ.

2.6 Perturbed plane waves

I am now in a position to describe the construction of analogues of the 'plane wave' eigenfunctions $\Phi_0(z, \omega, \lambda) = \exp(i\lambda z \cdot \omega)$ in the free case.

Lemma 2.4 *For each $0 \neq \lambda \in \mathbb{R}$ and $\omega \in \mathbb{S}^{n-1}$ there is a unique function in the null space of $\Delta + V - \lambda^2$ of the form*

(2.16)
$$\Phi_V(z, \omega, \lambda) = \exp(i\lambda z \cdot \omega) + \exp(-i\lambda|z|)|z|^{-\frac{1}{2}(n-1)}\phi_V(z, \omega, \lambda)$$
$$\text{with } \phi_V \in C^\infty(\mathbb{S}^n_+).$$

Proof The compactness of the support of V means that $(\Delta + V - \lambda^2)e^{i\lambda z \cdot \omega} = f$ with $f = -Ve^{i\lambda z \cdot \omega} \in C_c^\infty(\mathbb{R}^n)$. Since $R_V(\lambda)$ has no pole on $\mathbb{R} \setminus \{0\}$,

(2.17) $\Phi_V(z, \omega, \lambda) = \exp(i\lambda z \cdot \omega) + R_V(\lambda)f$

is, by (2.2) and Proposition 1.1, of the form (2.16) and satisfies $(\Delta + V - \lambda^2)\Phi_V = 0$. The uniqueness follows from Proposition 2.3, since the difference between two solution of the form (2.16) would be of the form (2.10).   □

Notice that $\Phi_V(z, \omega, \lambda)$, defined by (2.17), extends to be meromorphic as a function of λ on the same domain as $R_V(\lambda)$.

2.7 Relative scattering matrix

The coefficient in (2.16) also depends smoothly[16] on the parameters, i.e.

$$\phi_V(z, \omega, \lambda) \in C^\infty(\mathbb{S}^n_+ \times \mathbb{S}^{n-1} \times (\mathbb{R} \setminus \{0\})).$$

Thus, as with (1.9) in the free case, Φ_V can be used as the kernel of an operator into the null space of $\Delta + V - \lambda^2$:

---

[15] If V is complex-valued then it is possible for there to be poles of the resolvent at points in the continuous spectrum, but the corresponding eigenfunctions cannot be square-integrable.

[16] In fact ϕ_V is a real-analytic function of ω and λ on this set, see Footnote 26.

(2.18) $$\Phi_V(\lambda)g = \int_{\mathbb{S}^{n-1}} \Phi_V(z,\omega,\lambda)g(\omega)d\omega.$$

From the form of Φ_V in (2.17) and (1.13) it follows that $\Phi_V(\lambda)g$ has, for $g \in \mathcal{C}^\infty(\mathbb{S}^{n-1})$, a similar asymptotic expansion

(2.19)
$$\Phi_V(\lambda)g(|z|\theta) = e^{i\lambda|z|} (\lambda|z|)^{-\frac{1}{2}(n-1)} e^{-\frac{1}{4}\pi(n-1)i}(2\pi)^{\frac{1}{2}(n-1)}g(\theta)$$
$$+e^{-i\lambda|z|} (\lambda|z|)^{-\frac{1}{2}(n-1)} e^{\frac{1}{4}\pi(n-1)i}(2\pi)^{\frac{1}{2}(n-1)}g'(\theta) + u', \ u' \in L^2(\mathbb{R}^n).$$

In fact it has a complete asymptotic expansion of which I have written only the leading part. Thus for the perturbed problem $\Delta + V$, with $V \in \mathcal{C}_c^\infty(\mathbb{R}^n)$ real-valued, there is a complete analogue of Lemma 1.2, i.e. a solution of $(\Delta + V - \lambda^2)u = 0$ of the form (1.15) for each $h \in \mathcal{C}^\infty(\mathbb{S}^{n-1})$. Again from the uniqueness result, $h'(\theta)$ is determined by h. The map $A_V(\lambda)h = h'$ maps $\mathcal{C}^\infty(\mathbb{S}^{n-1})$ into itself. In fact, as follows from (2.19), it is of the form

(2.20) $$A_V h(\theta) = i^{(n-1)}h(-\theta) + \int_{\mathbb{S}^{n-1}} A_V'(\theta,\omega,\lambda)h(\omega)d\omega$$

where $A_V' \in \mathcal{C}^\infty(\mathbb{S}^{n-1} \times \mathbb{S}^{n-1} \times (\mathbb{R} \setminus \{0\}))$. The operator $A_V(\lambda)$ is the *absolute scattering matrix*. [17] From (2.16), (2.18) and (2.19), for $\lambda > 0$,

(2.21) $$A_V'(\theta,\omega,\lambda) = e^{\frac{1}{4}\pi(n-1)i}(2\pi)^{-\frac{1}{2}(n-1)}\lambda^{\frac{1}{2}(n-1)}\phi_V(\theta,\omega,\lambda).$$

As noted earlier it is usual to normalize away the scattering matrix for Euclidean space, and to define the scattering matrix[18] by

(2.22) $$S_V(\theta,\omega,\lambda)g = \left(i^{-(n-1)}A_V(\lambda)g\right)(-\theta).$$

Thus it follows that the scattering matrix is of the form $\mathrm{Id} + B_V$ where B_V is a smoothing operator, i.e. has $\mathcal{C}^\infty$ Schwartz' kernel. This kernel is called the scattering amplitude. Needless to say it carries the same information as A_V or S_V.

---

[17] I know, it is not a matrix, but this is by extrapolation from the 1-dimensional case where it is. Blame the physicists. The obvious alternative name, the 'scattering operator,' is reserved for the t-convolution operator with kernel obtained by Fourier transformation of the λ variable.

[18] I shall call this the relative scattering matrix.

2.8 Asymptotics of the resolvent

The generalized plane waves also give the leading term in the asymptotic expansion of $R_V(\lambda)f$, just as in Proposition 1.1 in the free case.

Proposition 2.4 *For* $f \in \dot{C}^\infty(\mathbb{S}^n_+)$ *and* $\lambda > 0$

(2.23)
$$R_V(\lambda)f = e^{-i\lambda|z|}|z|^{-\frac{1}{2}(n-1)}w(z),$$
$$\text{with } w = (SP^*)(w'), \ w' \in C^\infty(\mathbb{S}^n_+) \text{ and}$$
$$v = w' \upharpoonright \mathbb{S}^{n-1} = \frac{1}{2i\lambda}P^\dagger_V(\lambda)f = \frac{1}{2i\lambda}\int_{\mathbb{R}^n} P^\dagger_V(z', \frac{z}{|z|}, \lambda)f(z')dz'$$
$$\text{where } P_V(z';\omega,\lambda) = \lambda^{\frac{1}{2}(n-1)}e^{\frac{1}{4}\pi(n-1)i}(2\pi)^{-\frac{1}{2}(n-1)}\Phi_V(z,\omega,\lambda).$$

Proof To check (2.23) first notice that the boundary pairing, (2.13), whilst proved for two solutions as in (2.11), is also valid, with the same proof, when $u = u^{(1)} = R_V(\lambda)f$ is as in (2.11) but $u^{(2)} = \Phi_V(\lambda)$.[19] This gives the identity

(2.24) $\quad 2i\lambda\lambda^{-\frac{1}{2}(n-1)}e^{-\frac{\pi}{4}(n-1)i}(2\pi)^{\frac{1}{2}(n-1)}v = \int f(z)\Phi_V(z,\omega,\lambda)dz.$

Here v is the boundary value of $w' = R_V(\lambda)f$ in (2.2). This gives (2.23).
$\square$

As in the free case, $P_V(\lambda)$ is the Poisson operator, so that if $g \in C^\infty(\mathbb{S}^{n-1})$ then $u = P_V(\lambda)g$ is the unique solution of $(\Delta + V - \lambda^2)u = 0$ with an expansion as in (2.11) with $v_+ \upharpoonright \mathbb{S}^{n-1} = g$. Also, the difference of the resolvent at $\pm\lambda$ on the real axis can be recovered from the perturbed plane waves: [20]

(2.26)
$$R_V(\lambda) - R_V(-\lambda) = \lambda^{n-2}M_V(\lambda) = \frac{1}{2i}(2\pi)^{-(n-1)}\lambda^{n-2}\Phi_V(\lambda)\Phi^\dagger_V(-\lambda).$$

---

[19] Just integrate against a smooth function in the ω variable.

[20] Let u be the solution to $(\Delta + V - \lambda^2)u = 0$ obtained by applying $R_V(\lambda) - R_V(-\lambda)$ to $f \in C^\infty_c(\mathbb{R}^n)$. Thus u has an expansion as in (2.11) and v_+, the boundary value of w_+, which is just the boundary value of $-R_V(-\lambda)f$, is given by

(2.25) $\qquad \frac{1}{2i\lambda}\lambda^{\frac{1}{2}(n-1)}e^{-\frac{1}{4}\pi(n-1)i}(2\pi)^{-\frac{1}{2}(n-1)}\Phi^\dagger_V(-\lambda)f.$

The solution with this boundary value is obtained by applying $P_V(\lambda)$ and this is just the right side applied to f. It follows that the two sides of (2.26) are equal as operators.

2.9 L^2 eigenfunctions

The poles of the resolvent in the physical half-plane are the easiest to analyze. Indeed, since $R_V(\lambda)$ is holomorphic and acts on $\mathcal{S}(\mathbb{R}^n)$ in a deleted neighbourhood of such a point it follows that the residue operators[21] also act on $\mathcal{S}(\mathbb{R}^n)$. Their range is the corresponding eigenspace. Provided V is real, any poles in the physical half-plane are necessary simple and lie on the imaginary axis.[22] Conversely any L^2 eigenfunction of $\Delta + V$ corresponding to a negative eigenvalue, σ, gives rise to a pole of the resolvent $R_V(\lambda)$ at the unique point $\lambda \in \mathcal{P}$ with $\lambda^2 = \sigma$.

2.10 Zero energy states

All the L^2 eigenfunctions of $\Delta + V$ are known as 'bound states' in the physical literature. As well as those arising from negative eigenvalues there may be 'zero energy' states, i.e. elements of the null space of $\Delta + V$. For n odd, $R_V(\lambda)^{23}$ can have at most a double pole at 0.[24] The leading coefficient, i.e. the coefficient of λ^{-2}, maps onto the null space of $\Delta + V$ acting on $L^2(\mathbb{R}^n)$.[25] The residue space, being the linear span of the coefficients of λ^{-1} in the expansion of $R_V(\lambda)f$, for $f \in \mathcal{C}_c^\infty(\mathbb{R}^n)$ such that there is *no* λ^{-2} term, consists of elements of the null space of $\Delta + V$ which are not in $L^2(\mathbb{R}^n)$. For $n = 3$ this space is at most one

---

[21] Meaning the coefficients in the Laurent expansion of the operator near a pole $\widetilde{\lambda}$:

$$R_V(\lambda) = \sum_{j \leq p} (\lambda - \widetilde{\lambda})^{-j} B_j + B'(\lambda),$$

(2.27)

$B'(\lambda)$ holomorphic near $\widetilde{\lambda}$.

[22] That is, of order 1, i.e. $j = 1$ in (2.27). This follows from the self-adjointness, which is to say that for any $\phi \in \mathcal{S}(\mathbb{R}^n)$

(2.28)
$$\left| \int_{\mathbb{R}^n} \overline{\phi}(\Delta + V - \lambda^2)\phi \, dz \right| \geq 2|\operatorname{Im} \lambda||\operatorname{Re} \lambda|\|\phi\|_{L^2}^2$$
$$\implies \|\phi\|_{L^2} \leq (2|\operatorname{Im} \lambda||\operatorname{Re} \lambda|)^{-1}\|(\Delta + V - \lambda^2)\phi\|_{L^2}$$

[23] For V real-valued.
[24] This follows from the inequality (2.28).
[25] That it takes values in the L^2 null space of $\Delta + V$ follows from (2.28). To see that it is surjective observe that if $u \in L^2(\mathbb{R}^n)$ satisfies $(\Delta + V)u = 0$ then $\Delta u = -Vu \in \mathcal{C}_c^\infty(\mathbb{R}^n)$. This implies that $u = -R_0(0)Vu$ and that Vu cannot vanish identically so $(\operatorname{Id} - VR_0(\lambda))^{-1}Vu$ must have a double pole at 0, the leading coefficient of which is Vu. This shows that u occurs as the leading coefficient in $R_V(\lambda)u$.

dimensional and its non-zero elements are called 'half-bound states'; they do not occur if $n \geq 5$.

2.11 Meromorphy of the scattering matrix

The meromorphic continuation of the resolvent, to $\mathbb{C}$ for odd dimensions and to Λ for even dimensions, has been discussed above. From this it can be deduced that the scattering matrix has a similar continuation:

Proposition 2.5 [26] *The relative scattering matrix for* $V \in \mathcal{C}_c^\infty(\mathbb{R}^n)$ *extends to a meromorphic function of* $\lambda \in \mathbb{C}$ *for n odd and to a meromorphic function of* $\lambda \in \Lambda$ *for n even; the poles of the analytic continuation are among the poles of* $R_V(\lambda)$ *with at most the same multiplicity.*

---

[26] The Lipmann-Schwinger equation can be used to express the scattering matrix in terms of the resolvent. Thus, the kernel of the scattering matrix is, for $\lambda > 0$, the product of $\lambda^{\frac{1}{2}(n-1)} e^{\frac{1}{4}\pi(n-1)i} (2\pi)^{-\frac{1}{2}(n-1)}$ and the coefficient of $e^{-i\lambda|z|}|z|^{-\frac{1}{2}(n-1)}$ in the expansion of $\Phi_V(\lambda)$ as $|z| \to \infty$. Writing Φ_V as

(2.29) $\Phi_V(\lambda) = \Phi_0(\lambda) - R_0(\lambda)V\Phi_0(\lambda) + R_0(\lambda)VR_V(\lambda)V\Phi_0(\lambda)$

it follows that

(2.30) $A_V(\lambda) = A_0(\lambda) - \dfrac{1}{2i\lambda} P_0^\dagger \left(V\Phi_0(\lambda) - VR_V(\lambda)V\Phi_0(\lambda) \right),$ for $\lambda > 0.$

The compactness of the support of V means that the right side extends to a meromorphic function of λ, in $\mathbb{C}$ or Λ depending on the parity of the dimension. It follows that the poles of $A_V(\lambda)$ must occur at poles of $R_V(\lambda)$ and with no greater multiplicity.

3

Inverse scattering

In this lecture I shall discuss three main types of inverse scattering results. I shall ignore the case $n = 1$[1] since this has an enormous literature, the methods available then are somewhat different[2] and the results are more precise and complete. The three basic results examined here correspond to different parts of the scattering matrix. The first result, showing in particular that the scattering matrix $B_V(\theta, \omega, \lambda)$ determines V, arises from an examination of the high-frequency asymptotics, i.e. the behaviour as $\lambda \to \infty$, of B_V. The second result shows that, for $n \geq 3$, the scattering amplitude $B_V(\theta, \omega, \lambda)$ for fixed $0 \neq \lambda \in \mathbb{R}$ also determines V. Finally, in odd dimensions, the backscattering $B_V(-\omega; \omega, \lambda)$ defines a globally Fredholm map, so at least near most points small variations in the potential are determined by the scattering amplitude.

Before discussing these inverse results I shall briefly describe the Radon transform and then its modification to the 'Lax-Phillips transform,' which is the name I am giving to the explicit form of their 'translation-representation'[3] for the free problem.

3.1 Radon transform

Since it is has other applications below and is in any case the basis for the Lax-Phillips transform I shall first discuss the Radon transform. A

---

[1] This is a beautiful theory. A good starting point to investigate it is the work of Gel'fand and Levitan, [25].

[2] The basic difference is related to the possibility of complexification to a problem with one complex variable.

[3] I have rather underemphasized the Lax-Phillips approach to scattering theory, compared to its importance. This is partly because it cannot be applied without serious modification to many problems that still fit within the scope of scattering theory. Still it is an excellent paradigm. A brief discussion is given in Section 4.4.

hyperplane in $\mathbb{R}^n$ is determined by its unit normal vector, ω, and signed distance from the origin, s, where the sign is chosen so that $s\omega$ is a point in the hyperplane, a general point of which satisfies the linear equation $z \cdot \omega = s$. This parametrization is 2-1, since $(-s, -\omega)$ parametrizes the same hyperplane as (s, ω).

The Radon transform of $u \in C_c^\infty(\mathbb{R}^n)$ is the integral over hyperplanes:[4]

$$(3.1) \qquad Ru(s, \omega) = \int_{H=\{z \cdot \omega = s\}} u(z) dH_z \in C_c^\infty(\mathbb{R} \times \mathbb{S}^{n-1})$$

where dH_z is the Euclidean measure induced on H. Notice that $Ru(s, \omega) = 0$ in $|s| > R$ if $\operatorname{supp}(u) \subset B(R)$. Clearly $Ru(-s, -\omega) = Ru(s, \omega)$ so the map $u \mapsto Ru$ is not surjective.[5] That it is injective can be seen by relating it to the Fourier transform[6]:

$$(3.6) \qquad \widehat{u}(r\omega) = \int_{-\infty}^{\infty} e^{-irs} Ru(s, \omega) ds,$$

which is just the 1-dimensional Fourier transform in the s variable.

The Fourier inversion formula leads to the inversion formula for the

---

[4] There is an extensive theory of generalized Radon transforms in several different contexts. See Guillemin and Sternberg [30].

[5] In fact the Radon transform is not surjective as a map from $C_c^\infty(\mathbb{R}^n)$ to the even subspace of $C_c^\infty(\mathbb{R} \times \mathbb{S}^{n-1})$. The precise range is given by a result of Helgason [36]:

Theorem 3.1 *The range of the Radon transform as a map*

$$(3.2) \qquad R : \dot{C}^\infty(B(R)) \longrightarrow \dot{C}^\infty([-R, R] \times \mathbb{S}^{n-1}),$$

for any $R > 0$, consists of the functions satisfying

$$(3.3) \qquad g(-s, -\omega) = g(s, \omega) \ \forall \ s \in [-R, R], \ \omega \in \mathbb{S}^{n-1} \ and$$

$$(3.4) \qquad \int s^k g(s, \omega) ds = P_k(\omega) \ is \ a \ polynomial \ of \ degree \ at \ most \ k, \ \forall \ k.$$

[6] Just introduce polar coordinates $\xi = r\omega$ in the definition of the Fourier transform to see that

$$(3.5) \qquad \widehat{u}(r\omega) = \int e^{-ir\omega \cdot z} f(z) dz$$

from which (3.6) follows, since $dz = dH ds$ when ω is fixed.

Radon transform [7]

(3.10) $\qquad u(z) = \dfrac{1}{2}(2\pi)^{-n+1} \displaystyle\int_{\mathbb{S}^{n-1}} (|D_s|^{n-1} Ru)(z \cdot \omega, \omega) d\omega.$

This also gives Plancherel's formula

(3.11)
$$\int_{\mathbb{R}^n} |u(z)|^2 dz$$
$$= (2\pi)^{-n+1} \int_{\mathbb{R}} \int_{\mathbb{S}^{n-1}} ||D_s|^{\frac{1}{2}(n-1)} Ru(s, \omega)|^2 d\omega ds, \quad u \in \mathcal{C}_c^\infty(\mathbb{R}^n).$$

Notice that the Radon transform extends by continuity to a map from Schwartz space[8]

(3.12) $\qquad\qquad R : \mathcal{S}(\mathbb{R}^n) \longrightarrow \mathcal{S}(\mathbb{R} \times \mathbb{S}^{n-1}).$

There is a 'uniqueness' result for the Radon transform in this form which is convenient to invoke below, also due to Helgason.

Proposition 3.1 [9] *If $u \in \mathcal{S}(\mathbb{R}^n)$ has Radon transform with support in $[-R, R] \times \mathbb{S}^{n-1}$ then $u \in \dot{\mathcal{C}}^\infty(B(R))$.*

---

[7] The Fourier inversion formula written in polar coordinates is

(3.7) $\qquad\qquad u(z) = (2\pi)^{-n} \displaystyle\int_{\mathbb{S}^{n-1}} \int_0^\infty e^{iz \cdot r\omega} \widehat{u}(r\omega) r^{n-1} dr d\omega.$

Inserting (3.6) and using the fact that Ru is an even function to extend the integral over r to negative values gives

(3.8) $\quad u(z) = \dfrac{1}{2}(2\pi)^{-n+1} \displaystyle\int_{\mathbb{S}^{n-1}} \dfrac{1}{2\pi} \int_0^\infty e^{i(z \cdot \omega)r} \int_{-\infty}^\infty Ru(s, \omega) |r|^{n-1} ds \, dr \, d\omega.$

The 1-dimensional Fourier inversion formula then gives (3.10), with the operator on functions in s being $|D_s|^{n-1} = D_s^{n-1}$ if n is odd or

(3.9) $\qquad\qquad |D_s|^{n-1} v(s) = \dfrac{1}{2\pi} \displaystyle\int_{-\infty}^\infty e^{irs} \widehat{v}(r) |r|^{n-1} dr$

if n is even.

[8] The range space here consists of functions which are rapidly decreasing at infinity and remain so after any differential operator with constant coefficients in the first variable is applied.

[9] The characterization of the range on $\mathcal{S}(\mathbb{R}^n)$ follows as in Footnote 5 with condition (3.3) dropped, i.e. (3.4) holds for Ru whenever $u \in \mathcal{S}(\mathbb{R}^n)$. The injectivity of the Radon transform then shows that Proposition 3.1 follows from Theorem 3.1.

3.2 Wave group

The solution properties of the wave equation are fundamental to the
study of the high-frequency behaviour of the scattering amplitude, and
for other reasons besides.[10] This is encoded in the existence, and properties, of the wave group for the operator $\Delta + V$:[11]

Theorem 3.2 [12] *For any $V \in C_c^\infty(\mathbb{R}^n)$ and any $(u_0, u_1) \in C^\infty(\mathbb{R}^n)$ there
is a unique solution, $u \in C^\infty(\mathbb{R} \times \mathbb{R}^n)$ to the Cauchy problem*

$$\begin{aligned}
(D_t^2 - \Delta - V)u(t, z) &= 0 \text{ in } \mathbb{R}_t \times \mathbb{R}^n \\
u(0, z) &= u_0(z), \quad D_t u(0, z) = u_1(z) \ \forall \ z \in \mathbb{R}^n
\end{aligned}$$

(3.14)

and the resulting 2×2 matrix of operators

$$\text{(3.15)} \qquad U_V(t) \begin{pmatrix} u_0 \\ u_1 \end{pmatrix} = \begin{pmatrix} u(t, \cdot) \\ D_t u(t, \cdot) \end{pmatrix}$$

---

[10] In the more 'physical' literature Schrödinger's equation

(3.13) $(D_t + \Delta + V)v = 0$

is often used for similar purposes. The solution for the initial value problem
$v(0) = v_0$ is

$$v(t) = G(t)v_0 = \exp(-i(\Delta + V))v_0.$$

This defines a group of operators with infinitesimal generator $-i(\Delta + V)$, which
explains the otherwise enigmatic statement in Footnote 1.2.

[11] The discussion here is equally valid for the Laplacian of a Riemann metric on
a compact manifold (provided $|z - z'|$ is replaced by the Riemannian distance
between the points in (3.16)) with which the scattering case is compared at various
points below. The wave group for the Laplacian associated to a metric g will be
denoted $U_g(t)$.

[12] There are a variety of different approaches to the proof of this important result.
For the case, actually under discussion here, of a potential perturbation of the flat
metric the proof, as usual, can be accomplished by perturbation from the flat case
where the solution can be written down using the Fourier transform, or explicitly.
Another approach, developed systematically by John [47] and followed in [81], is
to use the Radon transform to write the solution as a superposition of perturbed
plane waves. This is particularly suited to the requirements of Section 3.5 below.
Another method due to Hadamard [33] (see [44] Chapter 17 or for a more complete
account the book of Friedlander [22]) allows the solution to be written down quite
explicitly. The theory of Fourier integral operators gives a representation of the
solution somewhat intermediate between these two formulations; it is often the
most convenient. The existence, uniqueness and support properties of the solution
can also be obtained by less explicit functional-analytic methods, based on energy
estimates.

forms a 1-parameter group[13] with infinitesimal generator $\begin{pmatrix} 0 & \mathrm{Id} \\ \Delta + V & 0 \end{pmatrix}$;
the solution to (3.14) also satisfies the support estimate [14]

(3.16)
$$\mathrm{supp}(u) \subset \{(t, z); \ \exists \ z' \in \mathrm{supp}(u_0) \cup \mathrm{supp}(u_1) \ with \ |t| \le |z - z'|\}.$$

One of the most important properties of the solution of the wave equation is the manner in which singularities of the solution of the initial value problem propagate. First note that the wave group extends by continuity[15] to a group of operators[16]

$$(3.17) \quad U_V(t) : C^{-\infty}(\mathbb{R}^n) \times C^{-\infty}(\mathbb{R}^n) \longrightarrow C^{-\infty}(\mathbb{R}^n) \times C^{-\infty}(\mathbb{R}^n)$$

with the same support property, (3.16). In fact the solution to (3.14) is an element of $C^\infty(\mathbb{R}_t; C^{-\infty}(\mathbb{R}^n))$.[17] In a weak form this can then be stated as a bound on the singular support of the solution:[18]

(3.18)
$$\mathrm{sing \ supp}(u) \subset \{(t, z) \in \mathbb{R} \times \mathbb{R}^n; \ \exists$$
$$z' \in \mathrm{sing \ supp}(u_0) \cup \mathrm{sing \ supp}(u_1) \ with \ |z - z'| = |t|\}.$$

In fact there is a much better way to describe such results, namely in terms of the wavefront set. [19] For a distribution on $\mathbb{R}^k$ this can be considered as a subset[20]

$$(3.19) \qquad\qquad \mathrm{WF}(u) \subset \mathbb{R}^k \times \mathbb{S}^{k-1}.$$

The wavefront set $\mathrm{WF}(u)$ consists of those points (z, ω) such that the Fourier transform of u localized near z is not rapidly decreasing in any

---

[13] That these operators form a group, $U_V(t) \cdot U_V(t') = U_V(t + t')$ follows from the uniqueness of the solution to (3.14) and the t-translation invariance of the problem.

[14] This is the 'finite speed of propagation' for solutions to the wave equation.

[15] In the topology of distributions.

[16] I denote by $C^{-\infty}(\mathbb{R}^n)$ the usual space of distributions of unrestricted growth, i.e. the dual space of $C_c^\infty(\mathbb{R}^n)$ denoted by Schwartz $\mathcal{D}'(\mathbb{R}^n)$.

[17] That is, it is a distribution in the spatial, z, variables which depends smoothly on the time variable t.

[18] That is, singularities travel only at the speed of 'light.' The singular support of a distribution is the complement of the largest (there is a largest) open set to which it restricts as a C^∞ function.

[19] Introduced in this C^∞ setting by Hörmander [40].

[20] The space here is to be thought of as the boundary of $\mathbb{R}^k \times \mathbb{S}_+^k$ where the second factor is really the stereographic compactification of the dual $\mathbb{R}^k$. Thus, and this is the more usual convention, $\mathrm{WF}(u)$ can be considered as a subset of $\mathbb{R}^k \times (\mathbb{R}^k \setminus \{0\})$ which is 'conic,' i.e. invariant under the $\mathbb{R}^+$ action on the second factor.

open cone containing the direction ω.[21] This notion allows (3.18) to be elegantly improved to the two statements[22]

(3.21) $\text{WF}(u) \subset \{((t,z),(\tau,\zeta)) \in \mathbb{R}^{n+1} \times \mathbb{S}^n ; \tau^2 = |\zeta|^2\}$ and

(3.22)
$$((t,z),(\tau,\zeta)) \in \text{WF}(u) \implies (z + t\frac{\zeta}{\tau}, \frac{\zeta}{|\zeta|}) \in \text{WF}(u_0) \cup \text{WF}(u_1).$$

The lines

(3.23) $t \longmapsto (t, z - t\frac{\zeta}{\tau}, \tau, \zeta),$

where $\tau^2 = |\zeta|^2$ as in (3.21), are the (null) bicharacteristics of the wave operator.[23] Then (3.22) results from the two statements that $\text{WF}(u)$ is a complete union of such bicharacteristic curves and that a bicharacteristic can only be in $\text{WF}(u)$ if the point on it over $t = 0$ arises from a point in $\text{WF}(u_0) \cup \text{WF}(u_1)$.

There is an intermediate space between the smooth functions of compact support and the distributions on which the wave group is bounded. Namely the finite energy space, $H_{\text{FE}}(\mathbb{R}^n)$, defined as the closure of $\mathcal{C}_c^\infty(\mathbb{R}^n) \times \mathcal{C}_c^\infty(\mathbb{R}^n)$ with respect to the norm

(3.24) $\|(u_0, u_1)\|^2 = \int_{\mathbb{R}^n} \left(|\zeta|^2 |\widehat{u}_0(\zeta)|^2 + |\widehat{u}_1(\zeta)|^2 \right) d\zeta$

This is a Hilbert space,

(3.25) $H_{\text{FE}}(\mathbb{R}^n) \subset \mathcal{S}'(\mathbb{R}^n) \times \mathcal{S}'(\mathbb{R}^n),$

on which $U_V(t)$ is a bounded group of operators.

---

[21] More precisely, the complement $\mathbb{R}^k \times \mathbb{S}^{k-1} \setminus \text{WF}(u)$, of the wavefront set of $u \in \mathcal{C}^{-\infty}(\mathbb{R}^k)$ is defined as consisting of those points (z, ω) such that for some function $\phi \in \mathcal{C}_c^\infty(\mathbb{R}^n)$, with $\phi(z) \neq 0$, and some function $\phi' \in \mathcal{C}^\infty(\mathbb{S}^{k-1})$ with $\phi'(\omega) \neq 0$ the product of ϕ' and the Fourier transform, $\phi'(\omega')(\widehat{\phi u})(r\omega')$ is uniformly rapidly decreasing as $r \to \infty$. This is an open condition, and hence $\text{WF}(u)$ is a closed set; its projection onto the first factor is the singular support.

(3.20) $\pi_{p,k} \text{WF}(u) = \text{sing supp}(u).$

See for example [42].

[22] This is almost a characterization of $\text{WF}(u)$ for the solution to (3.14). If $(z, \zeta) \in \text{WF}(u_0) \cup \text{WF}(u_1)$ then for at least one choice of sign, $((t, z \mp t\zeta), (\pm\frac{1}{\sqrt{2}}, \frac{1}{\sqrt{2}}\zeta)) \in \text{WF}(u).$

[23] That is, an integral curve of the Hamilton vector field $2\tau \partial/\partial t - 2\zeta \cdot \partial/\partial z$ of the symbol $\tau^2 - |\zeta|^2$, of the wave operator.

3.3 Wave operators

There is an alternate definition of the scattering matrix, indeed an alternative formulation of scattering theory itself, in terms of the wave group.[24] Namely, for $V \in \mathcal{C}_c^\infty(\mathbb{R}^n)$ real valued, the limits

$$(3.26) \quad W_\pm u = \lim_{t \to \pm\infty} U_V(-t)U_0(t)u, \ \forall \ u = (u_0, u_1) \in (\mathcal{C}_c^\infty(\mathbb{R}^n))^2$$

exist. This defines the Møller wave operators which extend by continuity to isomorphisms from the finite energy space onto the orthocomplement of the finite dimensional space spanned by the L^2 eigenfunctions, i.e. onto the part of the finite energy space corresponding to the continuous spectrum.[25] It is important that the two wave operators have the same range since this means that the 'abstract' scattering operator can be defined by:

$$(3.27) \qquad\qquad \widetilde{S}_V = W_+^{-1} \circ W_-.$$

It is then a unitary operator, provided V is real, acting on the finite energy space.[26]

3.4 Lax-Phillips transform

The Radon transform satisfies the identity[27]

$$(3.29) \qquad\qquad R\Delta f = D_s^2 R f$$

which effectively reduces the n-dimensional Laplacian to the one-dimensional Laplacian. This is the basis of the Lax-Phillips transform which,

---

[24] There is a similar formulation in terms of the group of operators solving the Schrödinger equation. That they give the same scattering operator (up to simple renormalization) is part of Kato's invariance principle [48]. For a development of scattering theory on this basis see the book of Yafaev [118] and the lectures of Simon [17].

[25] This is called the 'completeness' of the wave operators.

[26] The standard way of relating this scattering operator and the scattering matrix defined earlier is to observe, from (3.26), that $W_\pm U_V(t) = U_0(t)W_\pm$, i.e. the Møller wave operators intertwine the free and the perturbed wave group (acting on the orthocomplement of the bound states). This means that $\widetilde{S}_V$ commutes with $U_0(t)$ and so can be decomposed according to the spectral decomposition of the free Laplacian. As one would expect

$$(3.28) \quad \widetilde{S}_V \exp(i\lambda z \cdot \omega) = \exp(i\lambda z \cdot \omega) + \int_{s^{n-1}} B_V(\lambda)(\theta, \omega) \exp(i\lambda z \cdot \theta) d\theta.$$

[27] To obtain this just apply the Laplacian to the inversion formula, (3.10), and then apply the Radon transform.

for n odd, conjugates the free wave group $U_0(t)$ to a translation group. Define

(3.30)

$$\mathrm{LP}\begin{pmatrix} u_0 \\ u_1 \end{pmatrix} = (2\pi)^{-\frac{1}{2}(n-1)}\left(D_s^{\frac{1}{2}(n+1)}(Ru_0)(s,\omega) - D_s^{\frac{1}{2}(n-1)}(Ru_1)(s,\omega)\right).$$

The two terms, from u_0 and u_1, have opposite parities under $(s,\omega) \to (-s,\omega)$ so $\mathrm{LP} : \mathcal{C}_c^\infty(\mathbb{R}^n)^2 \longrightarrow \mathcal{C}_c^\infty(\mathbb{R} \times \mathbb{S}^{n-1})$ is injective.[28]

Proposition 3.2 *For any odd $n \geq 3$, the Lax-Phillips transform, (3.30), conjugates the free wave group to the translation group:*

(3.31) $\mathrm{LP} \circ U_0(t) \circ \mathrm{LP}^{-1} = W_0(t), \ \ W_0(t)k(s,\omega) = k(s - t,\omega).$

For any $V \in \dot{\mathcal{C}}^\infty(B(R))$ the transformed group[29]

(3.32) $W_V(t) = \mathrm{LP} \circ U_V(t) \circ \mathrm{LP}^{-1}$

satisfies

(3.33) $W_V(t)k(s,\omega) = k(s - t,\omega) \begin{cases} \textit{if } s \textit{ and } s - t < -R \\ \textit{or } s \textit{ and } s - t > R. \end{cases}$

3.5 Travelling waves

The spectral, and scattering, properties of $\Delta + V$ as $\lambda \to \infty$ can be expected to be classical, i.e. rather than involving global solutions of differential equations the asymptotic behaviour should be expressible in terms of local operations on V, and integration. To investigate the high-frequency behaviour I shall use the wave equation and Lax-Phillips transform.

The plane waves, $e^{i\lambda z\cdot\omega}$, on which much of the discussion up to this point has been based, can be obtained by the one-dimensional (inverse) Fourier transform:

(3.34) $e^{i\lambda z\cdot\omega} = \int_{\mathbb{R}} e^{-it\lambda}\delta(t + z \cdot \omega)dt.$

---

[28] In fact it extends by continuity to an isometric isomorphism from the finite energy space for the wave equation, defined in (3.25), onto $L^2(\mathbb{R} \times \mathbb{S}^{n-1})$; this indeed is the reason for the normalizing constant in (3.30).

[29] In the odd-dimensional case the infinitesimal generator of the transformed group $W_V(t)$ is $D_s + V_{\mathrm{LP}}$ where V_{LP} is the sum of a pseudodifferential operator of order -1 and and an 'anti-pseudodifferential' operator of order -1, by which I mean a pseudodifferential operator followed by the reflection $(s,\omega) \to (-s,-\omega)$. Both terms can be taken to have kernels with support in the region $|s|, |s'| \leq R$ if V has support in $|z| \leq R$.

Here $\delta(t + z \cdot \omega) \in \mathcal{S}'(\mathbb{R}^{n+1})$ is a distribution depending on $\omega \in \mathbb{S}^{n-1}$ as a parameter:

$$(3.35) \qquad \langle \delta(t + z \cdot \omega), \phi \rangle = \int_{\mathbb{R}^n} \phi(-z \cdot \omega, z) dz, \ \ \phi \in \mathcal{S}(\mathbb{R}^{n+1}).$$

By direct differentiation this distribution can be seen to be a solution of the (free) wave equation:

$$(3.36) \qquad\qquad (D_t^2 - \Delta)\delta(t + z \cdot \omega) = 0 \text{ in } \mathbb{R}^{n+1}.$$

If now a potential is added then a natural generalization is to seek a solution to the perturbed wave equation. Observe that the support of the 'travelling wave' $\delta(t + z \cdot \omega)$ is disjoint from the support of V when $t << 0$. Thus it makes sense to look for a perturbed travelling wave:

$$(3.37)$$
$$(D_t^2 - \Delta - V)u_V(t, z, \omega) = 0, \ \ u_V(t, z, \omega) = \delta(t + z \cdot \omega) \text{ in } t << 0.$$

This is the *forcing problem* for the wave equation.

Lemma 3.1 [30] *There is a unique solution to the forcing problem (3.37) and, for some $T > 0$, $\exp(-T(1 + t^2)^{\frac{1}{2}})u_V \in \mathcal{S}'(\mathbb{R}^n)$.*[31]

The forced behaviour of the solution in $t << 0$, together with (3.16), shows that for $|z| \leq R$ the solution u vanishes in $t < -C - R$ for some C depending on V. This, and the exponential bound, allows the Fourier-Laplace transform to be taken.

Proposition 3.3 [32] *For* $\text{Im}\,\lambda << 0$

$$(3.38) \qquad\qquad \Phi_V(z, \omega, \lambda) = \int_0^\infty e^{-it\lambda} u_V(t, z, \omega) dt$$

is the perturbed plane wave solution in Lemma 2.4.[33]

---

[30] The existence of the solution to this problem is equivalent to the solvability of the Cauchy problem; as discussed in Section 3.2 above. For a more leisurely treatment see [81].

[31] The only reason that the solution is not tempered is that there may be L^2 eigenfunctions.

[32] Certainly the Fourier-Laplace transform in (3.38) gives an element of the null space of $\Delta + V - \lambda^2$. That this solution is of the form (2.16) follows from the support properties of u_V.

[33] The analytic continuation of $\Phi_V(z, \omega, \lambda)$ being given by (2.17).

A solution to (3.37), modulo C^∞ errors, can be found in the form of a travelling wave with jump discontinuity:[34]

$$
(3.40) \quad \begin{aligned} u'_V(t,z,\omega) &= \delta(t + z \cdot \omega) + H(t + z \cdot \omega)b(t,z,\omega), \\ b &\in C^\infty(\mathbb{R}^{n+1} \times \mathbb{S}^{n-1}), \end{aligned}
$$

in the sense that $(D_t^2 - \Delta - V)u'_V \in C^\infty(\mathbb{R}^{n+1} \times \mathbb{S}^{n-1})$ and $b = 0$ in $t \ll 0$. The Taylor series of b at $t + z \cdot \omega = 0$ is uniquely determined by these conditions and

$$
(3.41) \quad b(-z \cdot \omega, z, \omega) = -\frac{1}{2}\int_0^\infty V(z + s\omega)ds.
$$

The solution u_V to the forcing problem is also of the form (3.40), with b satisfying (3.41).

3.6 Near-forward scattering

From the construction of u'_V as a solution, modulo C^∞, to the forcing problem the high-frequency asymptotics of the scattering amplitude $B_V(\theta, \omega, \lambda)$ can be deduced. As might be expected on geometric grounds, most of the scattering at high energy goes 'straight through' and in particular:

$$
(3.42) \quad \begin{aligned} &\text{For any } \epsilon > 0, \ N > 0 \ \exists \, C > 0 \text{ s.t.} \\ &|B_V(\theta, \omega, \lambda)| \le C(1 + |\lambda|)^{-N} \text{ if } |\theta - \omega| \ge \epsilon \text{ and } |\lambda| \ge \epsilon, \end{aligned}
$$

i.e. the scattering amplitude is rapidly decreasing as $\lambda \to \infty$ in any region where θ is bounded away from ω. On the other hand there is a 'peak' of energy rather narrowly focused near the forward direction $\theta = \omega$. Let $\omega^\perp \subset \mathbb{R}^n$ be the $(n-1)$-dimensional space of vectors orthogonal to the unit vector ω.

---

[34] Here $H(t)$ is the Heaviside function,

$$
(3.39) \quad H(t) = \begin{cases} 1 & t \ge 0 \\ 0 & t < 0 \end{cases}
$$

and b can be assumed to have support in $|t - z \cdot \omega| \le 1$.

Proposition 3.4 [35] *There are functions* $\alpha_j(\cdot;w) \in S(w^\perp)$, *for* $j = 0, \ldots, \infty$, *depending on* V *and smoothly on* $w \in \mathbb{S}^{n-1}$, *such that*

$$(3.43) \quad B_V(\theta, w, \lambda) \sim \sum_{j \geq 0} \lambda^{n-2-j} \alpha_j \big(\lambda(\theta - (\theta \cdot w)w); w\big) \ as \ \lambda \to \infty$$

with the leading term given by

$$\alpha_0(w; w) = c \int_{w^\perp} e^{i\xi \cdot w} \, \mathrm{T}_{XR}(V)(\xi, w), \ c \neq 0,$$

$$(3.44)$$

$$where \ \mathrm{T}_{XR}(V)(\xi, w) = \int_0^\infty V(\xi + sw) ds.$$

The function $\mathrm{T}_{XR}(V)$ is sometimes called the X-ray transform of V; it is just the integral over straight lines. The asymptotic expansion (3.43) shows that all the coefficients, in particular $\alpha_0(w; w)$ and hence $\mathrm{T}_{XR}(V)$, can be recovered from B_V.

Corollary 3.1 *The map* $V \longmapsto a_V$, *from (complex-valued) potentials in* $C_c^\infty(\mathbb{R}^n)$ *to the scattering amplitude, is injective.*

One way to see that V can be recovered from its X-ray transform, (3.44), is to use the Radon transform, since the Radon transform can be written in terms of the X-ray transform if $n \geq 2$. To do so, choose, for each $w \in \mathbb{S}^{n-1}$, a direction $\theta \in \mathbb{S}^{n-1}$ with $w \cdot \theta = 0$ and then[36]

$$(3.45) \qquad R(s, w) = \int_{\{w \cdot w = s, w \cdot \theta = 0\}} \int_{-\infty}^{\infty} V(w + t\theta) dt \, dw.$$

This formula, together with (3.10), shows that V can be recovered from α_0 in (3.44) and hence, by (3.43), from A_V.

In fact this procedure gives rather more, namely it gives a constructive procedure under which the potential can be recovered from the scattering amplitude.

3.7 Constant-energy inverse problem

Although this inverse result is quite satisfactory it does involve the high-frequency limit, which can be viewed as 'impractical.' The next result I

---

[35] This result can be obtained, for $n \geq 3$ odd, using the Lax-Phillips transform (3.38), (2.16) and (2.21). The even-dimensional case is similar but more involved.

[36] If $n = 2$ there is no integral over w here.

shall discuss only uses the restriction of scattering amplitude to a fixed
energy to recover the potential. It has the disadvantage, compared to
Corollary 3.1, that is is not constructive.[37] Note that this problem has
a considerable history. There are two main approaches, both of which
have succeeded.[38] Another interesting approach is through the work of
A. Melin, see for example [65].

Proposition 3.5 *For any $n \geq 3$ and any $0 \neq \lambda \in \mathbb{R}$ the map from
real potentials to scattering matrices at frequency λ, $C_c^\infty(\mathbb{R}^n) \ni V \longmapsto
A_V(\lambda) \in C^\infty(\mathbb{S}^{n-1} \times \mathbb{S}^{n-1})$ is injective.*

Suppose $V_i \in C_c^\infty(\mathbb{R}^n)$, $i = 1, 2$, are two real-valued potentials with the
same scattering matrix; I shall proceed to outline an argument showing
that $V_1 = V_2$. The first step is to consider $\Phi_i(z, \omega, \lambda) = \Phi_{V_i}(z, \omega, \lambda)$ for
$i = 1, 2$. Choose $R > 0$ so large that both V_1 and V_2 have support in the
interior of the ball $B(R)$ of radius R. For any functions $g_i \in C^\infty(\mathbb{S}^{n-1})$
put $u_i = \int_{\mathbb{S}^{n-1}} \Phi_i(z, \omega, \lambda) g_i(\omega) d\omega$, for $i = 1, 2$. Thus $(\Delta + V_i - \lambda^2)u_i = 0$.
It follows that

(3.46)

$$\int_{\mathbb{R}^n} (V_1(z) - V_2(z)) u_1(z) \overline{u_2(z)} dz$$

$$= -\int_{\mathbb{R}^n} \left((\Delta - \lambda^2)u_1 \cdot \overline{u_2} - u_1 \cdot \overline{(\Delta - \lambda^2)u_2} \right) dz$$

$$= 2i\lambda \int_{\mathbb{S}^{n-1}} \left(w_{+,1}\overline{w_{+,2}} - w_{-,1}\overline{w_{-,2}} \right) d\omega$$

[37] Although it can probably be made constructive.
[38] One approach is due to Fadeev, whose work was directed at generalizing the 1-
dimensional theory, in particular the original method of Gel'fand and Levitan.
A solution to the inverse scattering problem using this approach is described by
Henkin and Novikov in [37], see also the articles of Novikov [85] and [86], the
announcement [87] and the references therein; in particular this is dependent on
the work of Beals and Coifman [9] and Nachman [84]. The second approach, which
I shall follow, originates with Calderón [13], who considered a different inverse
problem, namely the recovery of the coefficients of a second-order operator from
the 'Dirichlet-to-Neumann' map on the boundary of a region, that map which
gives the normal derivative of the solution of the Dirichlet problem with given
data. Calderón's approach was extended and refined by Sylvester, Uhlmann and
others (see [111] and references therein). In particular Sylvester and Uhlmann
solved the inverse problem for the Dirichlet-to-Neumann map for $\Delta + V$ acting on
a ball.(See the survey of G. Uhlmann [113].) There is a close relationship between
these two problems to the extent that the results of [111] can be used to solve the
inverse scattering problem. In fact the argument I give here is a modified form
of Calderón's approach. Rather than proceeding via the Dirichlet-to-Neumann
problem (which is also used in [37]) this solves the inverse scattering problem
more directly, using the construction of exponential solutions in [111]. A more
leisurely discussion will appear in [81].

where $w_{\pm,i}$ are the boundary values of the u_i in the sense of Lemma 2.2. By definition of the scattering matrix $w_{-,1} = Aw_{+,1}$ and $w_{-,2} = Aw_{+,2}$ for the same operator A which is also unitary. Thus both sides in (3.46) vanish, i.e. for all choices of the g_i

(3.47)
$$\int_{B(R)} (V_1(z) - V_2(z))u_1(z)\overline{u_2(z)}dz = 0 \; \forall \; g_i \in \mathbb{S}^{n-1}, \; i = 1,2.$$

Lemma 3.2 [39] *For $V \in \mathcal{C}_c^\infty(\mathbb{R}^n)$ real-valued with support in $|z| < R$, and $0 \neq \lambda \in \mathbb{R}$, the functions $u = \int_{\mathbb{S}^{n-1}} \Phi_V(z,\omega,\lambda)g(\omega)d\omega$ are dense, in the topology of $\mathcal{C}^\infty(B(R))$, in the null space of $(\Delta + V - \lambda^2)$ acting on $\mathcal{C}^\infty(B(R))$.*

Thus from the assumption that V_1 and V_2 have the same scattering matrix at frequency λ it follows that

(3.48)
$$\int_{B(R)} (V_1(z) - V_2(z))u_1(z)u_2(z)dz = 0$$
$$\forall \; u_i \in \mathcal{C}^\infty(B(R)), \text{ with } (\Delta + V_i - \lambda^2)u_i = 0 \text{ for } i = 1,2.$$

3.8 Exponential solutions

To finish the proof of Proposition 3.5 it remains to construct enough solutions u_i as in (3.48) to show that $V_1 = V_2$. This amounts to showing that the linear span of the products $u_1 u_2$ of such solutions is dense in, say, $L^2(B(R))$. In fact it is convenient to show this through the use of the Fourier transform. Consider the set of $\zeta \in \mathbb{C}^n$ satisfying the two constraints[40]

(3.49)
$$\zeta \cdot \zeta = \lambda^2, \quad \text{Re}\,\zeta = \xi.$$

---

[39] The proof of this is a straightforward application of distribution theory. The dual space of $\mathcal{C}^\infty(B(R))$ is naturally identified with the space of all distributions on $\mathbb{R}^n$ with support in $B(R)$. Thus consider such a distribution f which vanishes when paired with all the u of the statement of the Lemma. This implies that

$$h(\omega) = \int_{\mathbb{R}^n} f(z)\Phi_V(z,\omega,\lambda)dz = 0 \; \forall \; \omega \in \mathbb{s}^{n-1}.$$

Here the integral is really a distributional pairing. The function h is actually the boundary value at infinity of $v = R_V(\lambda)f$. Thus it follows from the uniqueness theorem that v actually has support in $B(R)$. Since $f = (\Delta + V - \lambda^2)v$ it follows that $\int_{\mathbb{R}^n} f(z)u(z)dz = 0$ for any $u \in \mathcal{C}^\infty(B(R))$ satisfying $(\Delta + V - \lambda^2)u = 0$. This implies the density property of the Lemma.

[40] This is the part of the argument that originates with Calderón; see [13].

If ζ has the decomposition $\zeta = \xi + i\eta$ then the first condition requires $|\xi|^2 - |\eta|^2 = \lambda^2$ and $\xi \cdot \eta = 0$.

Lemma 3.3 [41] *For $V \in C_c^\infty(\mathbb{R}^n)$, real-valued with support in $|z| < R$, there exists $C > 0$ such that for each $\zeta \in \mathbb{C}^n$ satisfying (3.49) and $|\zeta| \geq C$, there is a solution $u_\zeta \in C^\infty(B(R))$ to $(\Delta + V - \lambda^2)u_\zeta = 0$ in $|z| \leq R$ of the form*

$$(3.50) \qquad u_\zeta = e^{iz \cdot \zeta}(1 + a(z, \zeta)) \text{ with } \sup_{|z| \leq R} |a(z, \zeta)| \leq C/|\zeta|.$$

If $n \geq 3$ then for each $\xi \in \mathbb{R}^n$ there exists $\xi' \in \mathbb{S}^{n-1}$ and $\xi'' \in \mathbb{S}^{n-1}$, with $|\xi'| = |\xi''| = 1$ and with all three mutually orthogonal. Then, for $t >> 0$

$$(3.51) \qquad \zeta_t^{\pm} = \xi \pm (t^2 - |\xi|^2 + \lambda^2)^{\frac{1}{2}}\xi' \pm it\xi''$$

satisfies (3.49). Let $u_1(t)$ be the solution, whose existence is guaranteed by Lemma 3.3, for ζ_t^+ and V_1. Similarly let $u_2(t)$ be the solution for ζ_t^- and V_2. As $t \to \infty$, $u_1(t)u_2(t) \to \exp(2iz \cdot \xi)$ on $B(R)$. Thus from (3.48) it follows that if $V = V_1 - V_2$ then

$$(3.52) \qquad \widehat{V}(-2\xi) - \lim_{t \to \infty} \int_{\mathbb{R}^n} (V_1 - V_2)u_1(t)u_2(t)dz = 0.$$

Since this is true for all ξ it follows that $V_1 = V_2$.

3.9 Backscattering

As I have already discussed, the scattering matrix does determine the potential V. A more refined form of this question is whether the 'backscattering' alone determines V. That is, whether V can be recovered from the restriction of the scattering matrix $B_V(\theta, \omega, \lambda)$ to the submanifold $\theta = -\omega$. This gives a formally determined problem in that the number of variables is again n. Eskin and Ralston [21] showed that the map from potential to backscattering is globally Fredholm. I shall give a form of this result which is easier to state; it is described in detail in [81]. This result only applies to potentials of compact support in odd dimensional spaces.

First observe that the function $B_V(\theta, -\theta, \lambda)$ can be obtained from the restriction of the scattering kernel, $k_V(s, \theta, -\theta)$. If V has support in $B(R)$ then $k_V(s, \theta, \omega)$ has support in $[-2R, \infty) \times \mathbb{S}^{n-1}$. This is always a

---

[41] The construction of these exponential solutions is close in spirit to the one-dimensional inverse methods. For details see [111].

smooth function, for $V \in C_c^\infty(B(R))$. Let $\mathcal{H}_{n,R} \subset \dot{H}^2([-2R, 2R] \times \mathbb{S}^{n-1})$ be the closure[42] of the image of the modified Radon transform:

$$(3.53) \quad C_c^\infty(B(2R)) \longmapsto D_s^{\frac{1}{2}(n-3)} RV \in \dot{H}^2([-2R, 2R] \times \mathbb{S}^{n-1}).$$

Then the modified backscattering transform

$$(3.54) \quad \begin{aligned} \beta : C^\infty(B(R)) \ni V &\longmapsto \\ \pi_R D_s^{\frac{1}{2}(n-3)} k_V(s, -\theta, \theta) &\in \mathcal{H}_{n,R} \subset \dot{H}^2([-2R, 2R] \times \mathbb{S}^{n-1}) \end{aligned}$$

is well defined, where π_R is the orthogonal projection onto $\mathcal{H}_{n,R}$.

Theorem 3.3 *For any odd $n \geq 3$, the backscattering transform (3.54) extends by continuity to a map*

$$(3.55) \quad \beta : \dot{H}^{\frac{1}{2}(n+1)}(B(R)) \longrightarrow \mathcal{H}_{n,R} \subset \dot{H}^2([2R, 2R] \times \mathbb{S}^{n-1})$$

which is globally analytic and Fredholm, i.e. β can be expanded in a convergent power series around each point $V \in \dot{H}^{\frac{1}{2}(n+1)}(B(R))$ and the derivative of β at any point is a Fredholm map.

It follows directly from this theorem that the map $V \to \beta(V)$ is locally an isomorphism near each point is some open dense set, which includes the zero potential. Whether or not β is everywhere a local isomorphism, or indeed is a global isomorphism, is not known.[43]

---

[42] Taken in $\dot{H}^2([-2R, 2R] \times \mathbb{S}^{n-1})$.
[43] Not known to me anyway.

4
Trace formulæ and scattering poles

The poles of the analytic continuation of the resolvent, $(\Delta + V - \lambda^2)^{-1}$, which arise from the L^2 eigenvalues, the half-bound states and the scattering poles (poles of the analytic continuation of the scattering matrix) are close analogues of the eigenvalues in the compact case. Indeed, in odd dimensions, they *are* the eigenvalues of an operator, namely the infinitesimal generator of the Lax-Phillips semigroup. The non-self-adjointness of this operator means that most of the techniques developed to analyze the eigenvalues of self-adjoint operators are not (directly) applicable. Nevertheless, many of the results from the spectral theory of the Laplacian on a compact manifold do have analogues for the scattering poles.

Often the results for compact problems are stated in terms of the counting function for the eigenvalues.[1] For scattering problems this has two natural, but different, analogues. The first is the scattering phase, the argument of the determinant of the relative scattering matrix.

---

[1] For the Laplacian of a Riemann metric, g, on a compact manifold without boundary the counting function for the eigenvalues can be taken to be

$$(4.1) \qquad N_g(\lambda) = \mathrm{sgn}(\lambda)\min\{j; \lambda_j > |\lambda|\}$$

where $0 = \lambda_0^2 < \lambda_1^2 \le \lambda_2^2 \le \ldots$ are the eigenvalues in increasing order repeated according to multiplicity. Thus, $N_g(\lambda)$ is automatically an odd function. The Poisson formula, the generalization to the scattering setting of which is discussed here, expresses the Fourier transform of the regularized trace of the wave group for the Laplacian (see [39] and [19]; note also Footnote 3.11)

$$(4.2) \qquad \tau_g(\rho) = \mathrm{Tr}\left(\int_{\mathbb{R}} U_g(t)\rho(t)dt\right), \ \forall \, \rho \in \mathcal{S}(\mathbb{R}) \Longrightarrow \tau_g \in \mathcal{S}'(\mathbb{R}),$$

in terms of the counting function:

$$(4.3) \qquad \widehat{\tau_g}(\lambda) = \frac{dN_g(\lambda)}{d\lambda} = \sum_{\lambda_j^2 \in \mathrm{spec}(\Delta_g)} \delta(\lambda - \lambda_j) + \delta(\lambda).$$

The first trace formula I will discuss expresses the relative trace of the wave group in terms of this scattering phase. In the odd-dimensional case there is a second trace formula, which is a more direct analogue of the Poisson formula. The relationship between these formulæ, and determinants, gives information about the distribution of the poles. For more detailed information the interested reader may also consult the recent survey article of Zworski, [121].

4.1 Determinant and scattering phase

The relative scattering matrix for a potential, $V \in \mathcal{C}_c^\infty(\mathbb{R}^n)$, is of the form $S_V(\lambda) = \mathrm{Id} + B_V(\lambda)$ with $B_V(\lambda)$ a smoothing operator acting on functions on the sphere at infinity. In particular, $B_V(\lambda)$ is trace class [2] so the Fredholm determinant

$$(4.5) \qquad d_V(\lambda) = \det S_V(\lambda), \ \lambda \in \mathbb{R} \setminus \{0\}$$

exists. If V is real valued then $S_V(\lambda)$ is unitary, so the determinant has absolute value 1. The scattering phase, defined as the argument:

$$(4.6) \qquad s_V(\lambda) = \frac{1}{i} \log d_V(\lambda) = \frac{1}{i} \log \det S_V(\lambda)$$

is therefore real. In fact (4.6) only fixes $s_V(\lambda)$ up to an additive constant in $2\pi\mathbb{Z}$.

For any $\rho \in \mathcal{C}_c^\infty(\mathbb{R})$ the time average of the difference of the perturbed and the free wave groups:

$$(4.7) \qquad U_V(\rho) - U_0(\rho) = \int_{\mathbb{R}} (U_V(t) - U_0(t)) \, \rho(t) dt$$

is a smoothing operator with compactly supported Schwartz kernel. It is therefore of trace class and its trace defines a distribution:

$$(4.8) \qquad \tau_V(\rho) = \mathrm{Tr} \int_{\mathbb{R}} (U_V(t) - U_0(t)) \, \rho(t) dt.$$

---

[2] For a discussion of the trace functional and trace class operators see Chapter 19 of [44], [27] or [99]. The latter includes a clear treatment of the Fredholm determinant. The Fredholm determinant can be defined by the condition (which is valid for matrices)

$$(4.4) \qquad \frac{d}{dt} \log \det(\mathrm{Id} + tA) = \mathrm{tr}\left(A(\mathrm{Id} + tA)^{-1}\right), \ \det \mathrm{Id} = 1$$

where A is in the trace class of operators. The right side of (4.4) is meromorphic with poles at the points $t = -1/\sigma_j$ with σ_j the non-zero eigenvalues of A; the residues are integers so $\det(\mathrm{Id} + tA)$ is entire in t.

Proposition 4.1 *For scattering by a potential the regularized trace of the wave group can be expressed in terms of the L^2 eigenvalues and the scattering amplitude in that*[3]

$$(4.9) \qquad \widetilde{\tau}_V(t) = \tau_V(t) - \sum_{(\lambda,k)\in D(V),\ \mathrm{Im}\,\lambda<0} 2k\cosh(i\lambda t)$$

is tempered and has Fourier transform

$$(4.10) \qquad \widehat{\widetilde{\tau}_V}(\lambda) = -i\,\mathrm{Tr}\left(S^{-1}(\lambda)\frac{d}{d\lambda}S(\lambda)\right).$$

The right side of (4.10) can also be interpreted in terms of the scattering phase so the complete, distributional, formula becomes

$$\tau_V(t) = \mathrm{Tr}\left(U_V(t) - U_0(t)\right)$$
$$(4.11)$$
$$= \sum_{\sigma\in\mathrm{ppspec}(\Delta+V)} 2\cosh(|\sigma|^{\frac{1}{2}}t) + \mathcal{F}^{-1}\frac{d}{d\lambda}s_V(\lambda).$$

As already noted the form of the singularities of the kernel of the wave group can be described rather precisely. This allows the regularity of $\tau_V(t)$, or equivalently $\widetilde{\tau}_V(t)$, to be analyzed:

Lemma 4.1 *If $\phi \in C_c^\infty(\mathbb{R})$ and $\phi - 1$ vanishes to infinite order at 0 then, provided[4] $n \geq 3$ is odd, $(1 - \phi(t))\widetilde{\tau}(t) \in \mathcal{S}(\mathbb{R})$. If n is even then $(1 - \phi(t))\widetilde{\tau}(t)$ has a complete asymptotic expansion as $|t| \to \infty$. There are constants w_j, $j \in \mathbb{N}$ such that for any $N \in \mathbb{N}$, $N > n/2$, as $t \downarrow 0$ if $n \geq 3$ is odd*

$$(4.12)$$
$$\tau(t) = \sum_{j=1}^{(n-1)/2} w_j D_t^{n-1-2j}\delta(t) + \sum_{j=(n+1)/2}^{N} w_j|t|^{2j-n} + r_N(t),$$
$$r_N \in C^{2N-n}(\mathbb{R})$$

whereas if $n \geq 2$ is even

$$(4.13)$$
$$\tau(t) = \sum_{j=1}^{n/2} w_j D_t^{2j-n}\log|t| + \sum_{j=n/2+1}^{N} w_j t^{n-2j}\log|t| + r_N(t),$$
$$r_N \in C^{2N-n}(\mathbb{R}).$$

---

[3] The divisor $D(V)$ is defined in Definition 2.1.
[4] For the 1 dimensional case see McKean and Singer [64].

The coefficients $w_j = w_j(V)$ are (essentially) the 'heat invariants' of the potential V. The first few of these can be written down explicitly:[5]

$$(4.14) \qquad w_1(V) = c_{1,n} \int V dz$$

$$(4.15) \qquad w_2(V) = c_{2,n} \int V^2 dz$$

$$(4.16) \qquad w_3(V) = c_{3,n} \int \left(V^3 + |\nabla V|^2 \right) dz$$

$$(4.17) \qquad w_4(V) = c_{4,n} \int \left(5V^4 + 5V^2 \Delta V + |\Delta V|^2 \right) dz;$$

the universal constants $c_{j,n}$ are all non-zero. In general, the jth invariant is the integral of a polynomial, with constant coefficients, in V and its derivatives where each term has weight j, with a factor of V considered to have weight 1 and each differentiation considered to have weight $\frac{1}{2}$. These polynomials are invariant under the action of $SO(n)$ on V.

From this short-time behaviour of the wave kernel it follows that $s(\lambda)$ has a complete asymptotic expansion as $\lambda \to \infty$:

$$(4.18) \qquad s_V(\lambda) \sim \sum_{j \geq 0} a_j(V) \lambda^{n-1-2j}.$$

The coefficients in this expansion are non-zero constant multiples of those in (4.12) or (4.13). Thus all the heat invariants $w_k(V)$ can be recovered from (the high-frequency asymptotics of) the determinant of the scattering matrix.

4.2 Poisson formula

The second and more subtle variant of the trace formula is only known in odd dimensions. It can be seen (very) formally by observing that the integral on the right in (4.10) is a contour integral which can be formally moved through all the poles of the scattering matrix to infinity. Although this deformation has not been justified directly the resulting formula can be proved using the Lax-Phillips semigroup.

---

[5] With some diligence these formulæ can be deduced from the results of [64].

Proposition 4.2 [6] *For scattering by $V \in C_c^\infty(\mathbb{R}^n)$ with $n \geq 3$ odd,*

$$(4.19) \qquad \mathrm{Tr}\,(U_V(t) - U_0(t)) = \sum_{(\lambda,k)\in D(V)} k e^{i\lambda|t|},\ t \neq 0,$$

where $D(V)$ is the divisor of V, see Definition 2.1.

I shall briefly discuss the proof of this result below.

4.3 Existence of poles

Combining this formula with (4.12) the existence of scattering poles follows for most potentials.

Proposition 4.3 [7] *For any odd dimension $n \geq 3$ and any real potential $V \in C_c^\infty(\mathbb{R}^n)$ such that $w_k(V) \neq 0$ for some $k \geq \frac{1}{2}(n+1)$, $D(V)$ is unbounded, i.e. the scattering matrix has an infinite number of poles.*

In particular, in view of the formula for $w_2(V)$ in (4.17):

Corollary 4.1 *For $n = 3$ there must be an infinite number of scattering poles for any potential of compact support which is not identically zero.*

Indeed this is to be expected in any dimension. The general properties of the invariants $w_k(V)$, together with Proposition 4.3, show that the set of potentials with an infinite number of scattering poles is open and dense in the C^∞ topology. For a fixed potential V there can be at most a finite set of real numbers, s, such that sV does not have an infinite number of scattering poles.

---

[6] This formula was proved first by Lax and Phillips [52] for obstacle scattering, but only for $t > 4R$ where the obstacle is contained in the ball of radius R. Bardos, Guillot and Ralston [6] extended the formula to the more natural range $t > 2R$. In [71] the trace formula is proved for all $t \neq 0$.

[7] This result does not seem to appear in the literature. The proof here occurred to me following discussions with Antonio Sá Barreto and Maciej Zworski, during the preparation of these notes, concerning their proof of such a result for potentials of a fixed sign (see [8]). The invariants $w_k(V)$ for $k \geq \frac{1}{2}(n+1)$ are, from (4.12), the coefficients of the powers $|t|^{2k-n}$ in the asymptotic expansion (really just a one-sided Taylor's formula) of $\tau_V(t)$ at $t = 0$. Thus if one of these is non-zero then the right side of (4.19) must be non-zero for t sufficiently small and positive. This shows that there must be at least one pole of the analytic continuation of the resolvent, so $D(V)$ is not empty. Suppose $D(V)$ were finite. Then the right side of (4.19) would be a finite sum. It would follow that the coefficient of t^0, i.e. the constant term, in the expansion of $\tau_V(t)$ as $t \downarrow 0$, would be the value at $t = 0$ of the sum, i.e. the number of elements of $D(V)$. However, from (4.12) the coefficient of t^0 vanishes in odd dimensions. Thus $D(V)$ cannot be both finite and non-empty. It follows that it is infinite, so the series on the right in (4.19) diverges at $t = 0$.

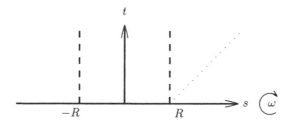

Fig. 6. The Lax-Phillips semigroup.

4.4 Lax-Phillips semigroup

In Section 3.4, I have already mentioned that the Lax-Phillips transform, for odd dimensions, gives an isomorphism of the finite energy space $H_{\mathrm{FE}}(\mathbb{R}^n)$ to $L^2(\mathbb{R} \times \mathbb{S}^{n-1})$ and intertwines the free wave group and the translation group. Moreover for a potential perturbation the wave group is conjugated to a group of bounded operators on $L^2(\mathbb{R} \times \mathbb{S}^{n-1})$ satisfying (3.33). This means that

(4.20)
$$k \in L^2([-R, R] \times \mathbb{S}^{n-1}) \subset L^2(\mathbb{R} \times \mathbb{S}^{n-1})$$
$$\implies W_V(t)k = 0 \text{ in } s < -R \text{ for } t > 0.$$

This allows the Lax-Phillips semigroup to be defined by orthogonal projection onto $L^2([-R, R] \times \mathbb{S}^{n-1})$[8]

(4.21)
$$Z_V(t)k = W_V(t)k \upharpoonright [-R, R] \times \mathbb{S}^{n-1}, \ k \in L^2([-R, R] \times \mathbb{S}^{n-1}), \ t \geq 0.$$

As the name suggests, this is a semigroup, $Z(t)Z(s) = Z(t+s)$, for $t, s \geq 0$. Its infinitesimal generator [9] is

(4.22)
$$Z_V(t) = \exp(itL_V), \ L_V = D_s + V_{\mathrm{LP}} \text{ with domain}$$
$$\mathrm{Dom}(L_V) = \big\{ k \in L^2([-R, R] \times \mathbb{S}^{n-1});$$
$$D_s k \in L^2([-R, R] \times \mathbb{S}^{n-1}) \text{ and } k(-R, \ ) = 0 \big\},$$

with V_{LP} as described in Footnote 3.29.

---

[8] That is, restriction back to $[-R, R] \times \mathrm{s}^{n-1}$.
[9] Existence of which is guaranteed by the Hille-Yosida theorem.

Lemma 4.2 *For any odd n and $V \in \dot{C}^\infty(B(R))$ the infinitesimal gen-*
erator of the Lax-Phillips semigroup has resolvent $(L_V - \lambda)^{-1}$, a mero-
morphic function of $\lambda \in \mathbb{C}$ with values in the bounded operators on
$L^2([-R, R] \times \mathbb{S}^{n-1})$ with poles of finite rank at precisely the poles of the
resolvent $R(\lambda)$ and with the same multiplicities.[10]

The proof of (4.19) uses this result. For $t > 2R$ the Lax-Phillips
semigroup $Z(t)$ is smoothing, and hence trace class as an operator on
$L^2([-R, R] \times \mathbb{S}^{n-1})$. Its trace is $\tau_V(t)$ and its non-zero eigenvalues are
$\exp(i\lambda t)$ for $\lambda \in D(V)$. In fact, for $T > 0$, a subspace of $L^2([-R, R] \times$
$\mathbb{S}^{n-1})$ can be constructed on which the operator $Z_V(t)$ has, for $t > T$,
the same eigenvalues as it has on $L^2([-R, R] \times \mathbb{S}^{n-1})$ and on which it is
of trace class. Lidskii's theorem, expressing the trace as the sum of the
eigenvalues, then gives (4.19).

I give another application of the Lax-Phillips semigroup in Section 4.6
below.

4.5 Counting function

Discussion of the convergence of the right side of (4.19), as with (4.3), is
related to counting functions for the poles. In the odd-dimensional case
I shall simply consider the counting function

$$(4.23) \qquad\qquad N(R) = \sum_{(\lambda,k) \in D(V), |\lambda| \leq R} k,$$

just the number of poles with absolute value less than R, as always
counted with multiplicity. The corresponding function for the case of
a compact manifold without boundary, see Footnote 1, has a leading
asymptotic term known as 'Weyl's law' which is[11]

(4.24)
$$N(R) = c_n \operatorname{Vol}(M)R^n + O(R^{n-1}), \quad n = \dim M, \text{ for } M \text{ compact.}$$

For scattering by a potential there is, in odd dimensions, a similar upper
bound.

Proposition 4.4 *If $n \geq 3$ is odd and $V \in C_c^\infty(\mathbb{R}^n)$ then*

$$(4.25) \qquad\qquad N(R) \leq C + CR^n.$$

---

[10] So giving the same divisor.
[11] See [19].

This result is due to Zworski, [119], and is shown to be optimal by an example in [120].[12] Zworski's argument to get the precise order (4.25), is a clever modification of the approach in [70]; it has been further simplified and extended by Vodev ([116]). Such bounds are obtained by showing that the poles are among the zeros of an entire function. As discussed in Lecture 2 the poles of the resolvent are the values of λ for which the Lipmann-Schwinger operator $\text{Id} + V R_0(\lambda)$ is not invertible as an operator on $L^2(B(R))$. Since $V R_0(\lambda)$ is compact this is the same as the values of λ for which -1 is an eigenvalue of $V R_0(\lambda)$. For $k \geq \frac{1}{2}(n+1)$ the product $(V R_0(\lambda))^k$ is trace class so the non-invertibility of $\text{Id} + V R_0(\lambda)$, which implies the non-invertibility of $\text{Id} + (V R_0(\lambda))^k$, means that the Fredholm determinant

$$(4.26) \qquad h_V^{(k)}(\lambda) = \det \left(\text{Id} + (V R_0(\lambda))^k \right)$$

must have a zero at that point. This remains true with multiplicities included, so to get an upper bound on the number of poles it suffices to count the zeros of the entire function $h_V^{(k)}(\lambda)$. For an entire function Jensen's inequality allows the counting function for the zeros to be bounded in terms of the growth of the function itself. In particular (4.25) follows from the estimate

$$(4.27) \qquad |h_V^{(n+1)}(\lambda)| \leq C e^{C|\lambda|^n}.$$

Estimates due to Weyl allow the determinant in (4.26) to be bounded in terms of the characteristic values of the operator $A = (V R_0(\lambda))^{n+1}$:

$$(4.28) \qquad |\det(\text{Id} + A)| \leq \prod_{j=0}^{\infty} (1 + \chi_j(A))$$

where the $\chi_j(A) > 0$ are positive square-roots of the eigenvalues of $A^* A$, repeated with multiplicity. If H_R is the characteristic function of a ball of radius R, containing the support of V then $V H_R = V$. The characteristic values of a product satisfy

$$(4.29) \qquad \begin{aligned} \chi_j(AB) &= \chi_j(BA), \\ \chi_{j+k}(AB) &\leq \chi_j(A)\chi_k(B) \text{ and} \\ \chi_j(AB) &\leq \|A\|\chi_j(B), \end{aligned}$$

so it suffices to consider the characteristic values of $H_R R_0(\lambda) H_R$, in which V no longer appears explicitly. In the physical half-space, $\text{Im}\,\lambda <$

---

[12] With n replaced by some integer this was proved in [70] and later improved to $n + 1$ in an unpublished note.

0, the characteristic values of $H_R R_0(\lambda) H_R$, are easily estimated uniformly and satisfy:

(4.30) $$\chi_j(H_R R_0(\lambda) H_R) \leq C j^{-\frac{1}{n}}, \ \operatorname{Im} \lambda \leq 0,$$

It follows that[13]

(4.31)
$$\chi_{j(n+1)}((V R_0(\lambda))^{n+1}) \leq C(\chi_j(H_R R_0(\lambda) H_R))^{n+1}$$
$$\Longrightarrow |h_V^{(n+1)}| \leq C \text{ in } \operatorname{Im} \lambda \leq 0.$$

To estimate the characteristic values of $H_R R_0(\lambda) H_R$ in $\operatorname{Im} \lambda \geq 0$ the identity (1.32) can be used. This allows the difference of the free resolvent at λ and $-\lambda$ to be written

(4.32) $$H_R \left(R_0(\lambda) - R_0(-\lambda) \right) H_R = i G_R(\lambda) G_R^*(\lambda)$$

where $G_R : L^2(\mathbb{S}^{n-1}) \longrightarrow L^2(B(R))$. The non-zero eigenvalues of $G_R G_R^*$ are the same as those of $G_R^* G_R$ which is an operator on $L^2(\mathbb{S}^{n-1})$. Thus it suffices to estimate, uniformly in λ, the characteristic values of the operator, Q_R, with Schwartz' kernel

(4.33) $$Q_R(\omega, \tau, \lambda) = \int_{B(R)} e^{i\lambda(\omega-\theta)\cdot z} dz \in C^\infty(\mathbb{S}^{n-1} \times \mathbb{S}^{n-1}).$$

This kernel is in fact real-analytic[14] and this leads to the uniform estimates

(4.35) $$\chi_j(H_R R_0(\lambda) H_R) \leq \begin{cases} C \exp 2T |\operatorname{Im} \lambda| & \text{for } j \leq C|\lambda|^{n-1} \\ C j^{-\frac{1}{n}} & \text{for } j > C|\lambda|^{n-1}. \end{cases}$$

This gives (4.27) for all $\lambda \in \mathbb{C}$ and hence (4.25).

As already noted, (4.25) is optimal, in the sense that there are examples with this growth. In general no lower bound is known.[15] The only general lower bound for potential scattering of which I am aware arises from the argument of Lax and Phillips for obstacle scattering [52], which has been be adapted by Vasy ([114]) to show

---

[13] For various constants, C.

[14] The norm, on $L^2(\mathbb{s}^{n-1})$, of $Q_{p,R} = (\Delta_{s^{n-1}} + 1)^p Q_R$ is, for some constant C which is independent of p and λ, bounded by $C^p(1 + |\lambda|^{2p}) \exp(2R|\operatorname{Im} \lambda|)$. Since $Q_R = (\Delta_{s^{n-1}} + 1)^{-p} Q_{p,R}$ this leads to bounds on the characteristic values in terms of the eigenvalues of the Laplacian on the sphere:

(4.34) $$\chi_j(Q_R) \leq j^{-p\frac{1}{n-1}} (C')^p (1 + |\lambda|^{2p}) \exp(2R|\operatorname{Im} \lambda|).$$

Taking the minimum over p of the right side, for each fixed j gives (4.35).

[15] Corollary 4.1 shows that for $n = 3$ and any potential, other than 0, there is an infinite set of poles. The corresponding result is not known to hold in odd dimensions $n \geq 5$, or in any even dimension.

Lemma 4.3 *For $n = 3$ and any potential $V \in C_c^\infty(\mathbb{R}^n)$ of fixed sign, not identically zero, the number of pure-imaginary poles satisfies*

$$(4.36) \qquad N_{\mathrm{Im}}(R) = \sum_{(is,k) \in D(V), s \in [-R,R]} k \geq cR^{n-1}, \ c > 0.$$

For the even-dimensional case bounds on the number of poles have been given by [46] and improved by Vodev ([117]) to

$$(4.37) \qquad N(R,a) = \sum_{\substack{(\lambda,k) \in D(V), |\lambda| \leq R, \\ -a \leq \arg \lambda \leq a}} k$$

$$N(R,a) \leq Ca\left(R^n + (\log a)^n\right), \ a, R > 1$$

where, as usual, $\lambda \in \Lambda$ is described formally by extension of the argument to $-\infty < \arg \lambda < \infty$.

4.6 Pole-free regions

There are only a finite number of poles of the analytic continuation of the resolvent in the physical region, $\mathrm{Im} \, \lambda \leq 0$. In the odd-dimensional case this can be extended to give pole free regions, near infinity, in which $\mathrm{Im} \, \lambda \to \infty$:

Proposition 4.5 *For $n \geq 3$ odd and any $V \in C_c^\infty(\mathbb{R}^n)$, real-valued, and any constant N there are only finitely many poles of the resolvent satisfying*

$$(4.38) \qquad \mathrm{Im} \, \lambda \leq N + N \log(1 + |\mathrm{Re} \, \lambda|).$$

Results of this type were obtained by Lax and Phillips, [53], but for a fixed N. The general estimate, (4.38), follows from the fact that for $t > 2R$ the Lax-Phillips semigroup, $Z_V(t)$, has a C^∞ kernel as an operator on $[-R, R] \times \mathbb{S}^{n-1}$ and depends smoothly on t. Thus the t-derivative of any order, $D_t^p Z(t)$, is bounded for any p. The eigenvalues are therefore bounded and thus

$$(4.39) \qquad |\lambda_j|^p \exp(-T \, \mathrm{Im} \, \lambda_j) \leq C_{p,T}$$

holds for all j. This implies (4.38).

5

Obstacle scattering

So far I have considered, in the main, scattering theory for potential perturbations of the flat Laplacian. In this lecture I wish to discuss some of the properties of the more substantial perturbation obtained by introducing a boundary in a compact region; in the last three lectures I shall consider various problems which are not perturbations of Euclidian space at all. Even restricting attention to compactly-supported perturbations of Euclidian space, one can consider much more general situations. For example the metric can be changed on a compact set,[1] the topology itself can be modified or the Laplacian can be deformed in more serious ways, say to a hypoelliptic operator.[2] Rather generally one can consider a 'black box' perturbation[3] with appropriate analytic properties, and still variants of many of the results discussed below remain valid. Nevertheless it is worthwhile considering the different types of perturbations separately, since they each have their own special features. Even the subject of scattering by an obstacle is a large, and in parts quite technical, one to which I cannot do justice in one lecture. I shall therefore limit myself to discussing some representative results which indicate both the similarities to, and differences from, the case of potential scattering.

---

[1] Such perturbations are included as a 'trivial case' in the discussion in Lecture 6.

[2] For example it is only in cases like this, where the 'spectral density' is higher for the perturbation than for the free problem that a leading asymptotic term for the counting functions for the poles has been shown to exist – see [109] and for related results [115] and [89].

[3] For example, the analytic continuation of the resolvent for such general perturbations is discussed by Sjöstrand and Zworski in [108].

5.1 Obstacles

Consider a smooth 'obstacle.' This can be specified by a smooth compact embedded (but not necessarily connected) hypersurface $H \subset \mathbb{R}^n$ such that

(5.1) $\mathbb{R}^n \setminus H = \Omega \cup \mathcal{O}$, with $\overline{\mathcal{O}}$ compact and Ω connected

and with both $\mathcal{O}$ and Ω open. Thus $\overline{\mathcal{O}} = \mathcal{O} \cup H$ is the 'obstacle' and Ω is its exterior. On the boundary, H, of Ω a self-adjoint boundary condition, such as the Dirichlet, Neumann or Robin condition, should be specified for the Laplacian; for the sake of definiteness I shall only consider the Dirichlet condition.

Using methods similar to those already described in the case of a potential it follows that:[4]

Proposition 5.1 *For each $0 \neq \lambda \in \mathbb{R}$ and each $\omega \in \mathbb{S}^{n-1}$ there is a unique function $u \in \mathcal{C}^\infty(\overline{\Omega})$ satisfying $(\Delta - \lambda^2)u = 0$ in Ω and such that*

(5.2)
$$u = e^{i\lambda z \cdot \omega} + u', \ u' = e^{-i\lambda|z|}|z|^{-\frac{1}{2}(n-1)}g(\frac{z}{|z|}) + u'', \ u'' \in L^2(\Omega)$$

where $u = 0$ on H and $g \in \mathcal{C}^\infty(\mathbb{S}^{n-1})$.

This can be seen from the analytic continuation of the resolvent $(\Delta - \lambda^2)^{-1}$. As for potential scattering this operator, which is bounded on $L^2(\Omega)$ for λ in the physical space, the lower half-plane, continues analytically to an operator mapping $\mathcal{C}_c^\infty(\overline{\Omega})$ to $\mathcal{C}^\infty(\overline{\Omega})$ and depending meromorphically on $\lambda \in \mathbb{C}$ or $\lambda \in \Lambda$ according to the parity of the dimension.[5] For scattering by an obstacle there are no bound states; the only spectrum is $[0, \infty)$ with no embedded eigenvalues. [6]

---

[4] This result, for odd dimensions, is contained, for example, in the book of Lax and Phillips, [51].

[5] The general strategy for proving such a result is to find a parametrix, i.e. an inverse for the problem (which should be holomorphic in $\lambda \in \mathbb{C}$ or Λ, or at least an arbitrarily large open subset), up to a compact error term. For potential scattering the resolvent of the free Laplacian gives such a parametrix. For an obstacle problem the theory of elliptic boundary problems (see for example [44]) gives a local parametrix for the boundary problem. This can be combined, using a partition of unity, with the free resolvent near infinity to give a global parametrix. The assumptions on a 'black box' perturbation should be made so that this type of construction can be applied, see [108]. Once the parametrix has been constructed arguments much as for potential scattering apply. For instance the resolvent of the boundary problem can be related to that of the free problem, enough to give Proposition 5.1.

[6] The proof of the absence of embedded eigenvalues follows as in the potential case.

Again as in the case of potential scattering it follows that, given $0 \neq \lambda \in \mathbb{R}$, for each $g \in \mathcal{S}(\mathbb{R}^n)$ and $f \in \mathcal{C}^\infty(\mathbb{S}^{n-1})$ there is a unique solution $u \in \mathcal{C}^\infty(\Omega)$ of

(5.3)
$$(\Delta - \lambda^2)u = g \text{ in } \Omega \text{ satisfying } v = 0 \text{ on } H \text{ and}$$
$$v = e^{i\lambda|z|}|z|^{-\frac{1}{2}(n-1)}f(\frac{z}{|z|}) + e^{-i\lambda|z|}|z|^{-\frac{1}{2}(n-1)}f_-(\frac{z}{|z|}) + v', \quad v' \in L^2(\Omega)$$

for some $f_- \in \mathcal{C}^\infty(\mathbb{S}^{n-1})$. The generalized plane waves in Proposition 5.1 give a parametrization of the null space of the Laplacian on tempered distributions (satisfying the boundary condition in a weak sense). The solutions in (5.3) are dense in this space, and in fact there is a generalization where $f \in \mathcal{C}^{-\infty}(\mathbb{S}^{n-1})$ which generates the whole null space. If $g \equiv 0$ then f_- is determined by f and the map

(5.4)
$$A_H(\lambda) : f \longmapsto f_-$$

is the *absolute scattering matrix*. From (5.2) and (1.13) it follows that

(5.5)
$$A_H(\lambda)f(\theta) = i^{(n-1)}f(-\theta) + (A'_H f)(-\theta)$$

with A'_H a smoothing operator, i.e. of the form

(5.6)
$$A'_H f(\theta) = \int_{\mathbb{S}^{n-1}} A'_H(\theta, \omega, \lambda)f(\omega)d\omega,$$

the *(relative) scattering matrix* $A'_H(\theta, \omega, \lambda)$ being a smooth function of $\theta, \omega \in \mathbb{S}^{n-1}$ for each $0 \neq \lambda \in \mathbb{R}$.[7] Again as in the potential case[8]

(5.7)
$$f \text{ and } f_+ \equiv 0 \text{ in } (5.3) \Longrightarrow v \in \mathcal{S}(\overline{\Omega});$$
$$f \text{ and } f_+ \equiv 0 \text{ and } g \in \mathcal{C}_c^\infty(\mathbb{R}^n) \Longrightarrow v \in \mathcal{C}_c^\infty(\overline{\Omega}).$$

It is also the case that no two different obstacles can have the same scattering matrix.

Proposition 5.2 [9] *If $n \geq 3$ is odd and H_i, $i = 1, 2$ are two smooth*

That there are no bound states is a consequence of the formal positivity of the Laplacian with Dirichlet (or Neumann) boundary condition.

[7] In fact it is real-analytic.

[8] Indeed the proof reduces to the free case. Choose a function $\phi \in \mathcal{C}_c^\infty(\mathbb{R}^n)$ which is equal to 1 in a ball containing H in its interior. Then consider $v' = (1 - \phi)v$; it satisfies $(\Delta - \lambda^2)v' \in \mathcal{S}(\mathbb{R}^n)$ and has the same asymptotic expansion as in (5.3). Thus $f \equiv f_+ \equiv 0$ implies $v' \in \mathcal{S}(\mathbb{R}^n)$ and hence $v \in \mathcal{S}(\mathbb{R}^n)$. Similarly if $g \in \mathcal{C}_c^\infty(\mathbb{R}^n)$ as well then v' must have compact support.

[9] This result can be found as Theorem 5.6 of Chapter 5 of [51], where the proof is attributed to Schiffer. For the given $f \in \mathcal{C}^\infty(\mathbb{s}^{n-1})$ consider the solutions v_i, $i = 1, 2$, to (5.3) with $g \equiv 0$ for the two obstacles with boundaries H_i. Using a

embedded hypersurfaces as above then $A_{H_1}f = A_{H_2}f$ *for any one* $f \in$ $C^\infty(\mathbb{S}^{n-1})$,[10] *not identically zero, for all* $0 \neq \lambda \in \mathbb{R}$ *implies* $H_1 = H_2$.

5.2 Scattering operator

Proposition 5.2 is an (older) analogue of Proposition 3.5 and is similarly non-constructive, i.e. does not give a procedure by which H can be recovered from A'_H. In search of a more constructive recovery procedure consider the *absolute scattering kernel*

$$(5.8) \qquad B_H(t,\theta,\omega) = \frac{1}{2\pi}\int_{\mathbb{R}} e^{-i\lambda t} A_H(\theta,\omega,\lambda)d\lambda$$

and the corresponding *relative scattering kernel* which is given by

$$B'_H(t,\theta,\omega) = i^{n-1}\left(B_V(t,-\theta,\omega) - \delta(t)\delta_\omega(\omega)\right).$$

These are both distributions in the space $C^{-\infty}(\mathbb{R}\times\mathbb{S}^{n-1}\times\mathbb{S}^{n-1})$ with the support property

$$(5.9) \qquad H \subset \{|z| \leq R\} \Longrightarrow \operatorname{supp}(B'_H) \subset \{t \geq -2R\}.$$

The scattering operator is the operator on $\mathbb{R}\times\mathbb{S}^{n-1}$ defined by B_H, or $\delta(t)\delta_\omega(\theta) + B'_H$ acting as a convolution operators in the t variable. The most immediate invariants of such an operator are the singularities of the kernel.

For the case of potential scattering it follows from the construction of perturbed travelling wave solutions to the wave equation that

$$(5.10) \qquad \operatorname{sing\ supp}(B'_V) \subset \{t = 0, \theta = \omega\}.$$

In fact more is true. Not only is the singular support of B'_V contained

---

cutoff as in Footnote 8 it follows that $v_1 - v_2$ has compact support. By unique continuation for elements of the null space of $\Delta - \lambda^2$ it follows that $v_1 - v_2$ must vanish on the complement of $\mathcal{O}_1 \cup \mathcal{O}_2$ and hence on its boundary. If $H_1 \neq H_2$ then, by relabelling it can be supposed that $\Omega_2 \setminus \mathcal{O}_1 \neq \emptyset$. Consider any component of $\Omega_2 \setminus \mathcal{O}_1$. The boundary of such an open set is contained in the union of the boundaries of $\mathcal{O}_1$ and of the boundary of $\mathcal{O}_1 \cup \mathcal{O}_2$. In either case v_2 must vanish there, so v_2 is an eigenfunction, with Dirichlet boundary condition in this set. Since the Dirichlet spectrum on any bounded open set has a positive lower bound, it follows that for λ small enough $v_2 \equiv 0$ in such an open set. However v_2 is real-analytic in Ω_2 so this implies $v_2 \equiv 0$ in Ω_2. This contradiction implies that $H_1 = H_2$.

[10] The same argument works if $f \in C^{-\infty}(\mathbb{s}^{n-1})$ and in particular if f is taken to be the delta measure at some point ω then the identity of the two scattering matrices for all $\theta \in \mathbb{s}^{n-1}$ and for all λ and one ω implies that $H_1 = H_2$.

in the 'diagonal' but, acting as a convolution operator in t :

$$(5.11) \qquad B'_V * f(t, \theta) = \int_{-\infty}^{\infty} \int_{\mathbb{S}^{n-1}} B'_V(t - t', \theta, \omega) f(t', \omega) d\omega dt',$$

B'_V is a pseudodifferential operator. More precisely, the wavefront set of the scattering kernel satisfies

(5.12)
$$\mathrm{WF}(B'_V) \subset \{(0, \omega, \omega; \tau, \tau w, -\tau w) \in T^*(\mathbb{R} \times \mathbb{S}^{n-1} \times \mathbb{S}^{n-1});$$
$$\tau \neq 0 \text{ and } \exists z \in \mathrm{supp}(V) \text{ with } (z - w) \perp \omega\}.$$

Thus the singularities of B'_V can be attributed to the geodesics which pass through the support of V. Now, a geodesic in $\mathbb{R}^n$ is a curve of the form

$$(5.13) \qquad z(t) = w' + (t - t')\omega', \quad \omega' \in \mathbb{S}^{n-1}$$

where w' and ω' are uniquely determined if it is required that $w' \cdot \omega' = 0$. The geodesic passes through $\mathrm{supp}(V)$ if $z = w' + t''\omega' \in \mathrm{supp}(V)$ for some t'', i.e. w' is in the projection of $\mathrm{supp}(V)$ into the plane orthogonal to ω. For each $0 \neq \tau \in \mathbb{R}$, this gives a point in the set on the right in (5.12) and every such point arises this way.

5.3 Reflected geodesics

To generalize (5.12) to obstacle scattering the notion of a reflected geodesic is needed. I shall consider only '$\mathcal{C}^\infty$' geodesics,[11] those which are related to the propagation of singularities in the $\mathcal{C}^\infty$ sense. A reflected geodesic is a curve $I \ni t \longrightarrow z(t) \in \overline{\Omega}$, where $I \subset \mathbb{R}$ is an interval, which satisfies a condition near each point, depending on the nature of the point. A curve is admitted as a reflected geodesic if it satisfies one of the following conditions near each point $z(t')$:[12]

(i) $z(t') \in \Omega$ and $z(t)$ is a free geodesic nearby.

Thus $z(t)$ is in the interior nearby and the curve is a straight line segment as in (5.13), for $|t - t'| < \epsilon$ for some $\epsilon > 0$.

(ii) $z(t') \in H$ with transversal, equal-angle reflection.

---

[11] Note that the '$\mathcal{C}^\infty$' definitely refers to the space of functions modulo which the singularities travelling along the curves are computed, not the curves themselves which are not even C^1 in general.

[12] Of course $t \in I$ throughout.

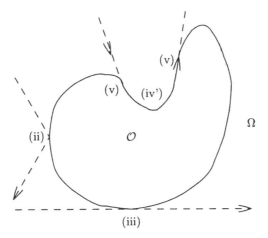

Fig. 7. Reflected geodesics.

By this I mean that if ν is the unit inward-pointing (with respect to Ω) normal to H at $z(t')$ then there are two directions ω', $\omega'' \in \mathbb{S}^{n-1}$ such that $\omega'' - \omega' = c\nu$, $c > 0$, and for some $\epsilon > 0$, $z(t) = z(t') + (t - t')\omega'$, for $t' - \epsilon < t < t'$, and $z(t) = z(t') + (t - t')\omega''$ for $t' < t < t' + \epsilon$. This is Snell's law.

(iii) $z(t') \in H$ and $z(t)$ is a diffractive ray nearby.

This means that for some $\omega \in \mathbb{S}^{n-1}$ with $\omega \cdot \nu = 0$, and some $\epsilon > 0$, $z(t) = z(t') + (t - t')\omega$ where the ω-directional curvature of H, with respect to Ω is strictly negative at $z(t')$.[13]

(iv) $z(t') \in H$, $z(t) - z(t') - (t - t')\omega = O(|t - t'|^{\frac{3}{2}})$ with $\omega \cdot \nu = 0$ and the ω-directional curvature is non-negative at $z(t')$.

This is the residual case and can naturally be divided into two. If the ω-directional curvature is strictly positive at $z(t')$ then it is necessarily positive nearby. It can then be shown that

(iv') $z(t') \in H$ and $z(t)$ is a gliding ray nearby.

This means that $z(t) \in H$ for $|t - t'| < \epsilon$ and $z(t)$ is a geodesic for the induced Riemann structure on H. This leaves only the cases of 'higher order contact':

---

[13] The directional curvature is the curvature of the curve formed by $H \cap P$ as a curve in the 2-plane $P = \mathrm{sp}\{x\nu + y\omega\}$ where ν is the unit normal as before.

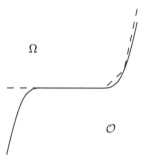

Fig. 8. Non-uniqueness of extension of reflected geodesics.

(v) $z(t') \in H$, $z(t) - z(t') - (t - t')\omega = O(|t - t'|^{\frac{3}{2}})$ with $\omega \cdot \nu = 0$
and the ω-directional curvature is zero at $z(t')$.

As with geodesics on a complete manifold, such as $\mathbb{R}^n$, these reflected
geodesics can be continued indefinitely in each direction. However, in
general there is no uniqueness for the continuation. Non-uniqueness can
occur only at points at which the directional curvature vanishes to infinite order. If the boundary H is real-analytic there is always uniqueness.
The following example of non-uniqueness of the continuation of rays is
due to Taylor ([112]).

Example Consider a region in the plane bounded by a curve on which
there is a point at which the curvature vanishes to infinite order but at
which it has a simple change of sign, i.e. it is negative on one side and
positive on the other. There is then a straightline segment in Ω which is
tangent to the boundary, to infinite order, at the flat point; see Figure 8.
This segment lies on the side on which the obstacle is locally convex. It
is possible to arrange that, on the side of the curve on which the obstacle
is concave, there is a ray which is reflected infinitely often with a limit
point at the point of vanishing curvature.[14] It is straightforward to
check that, together, the free segment and the infinitely reflected curve
form a reflected ray according to the definition above. However, it is

---

[14] Draw in the infinitely reflected ray first by taking a sequence of intervals, the sum
of the lengths of which converges, and arrange them end to end with successive
angle changes which are all of fixed sign and are small compared to the lengths
(i.e. the angles tend to zero when divided by any fixed power of the length). Then
it is easy to show that a curve can be drawn through the points of reflection so
that Snell's law holds and the curvature has the stated properties.

also the case that the free segment together with the boundary curve in the concave region constitutes a reflected ray. This shows that there is non-uniqueness for the local continuation of the incoming free segment, as a ray.[15]

5.4 Ray relation

For each $\omega \in \mathbb{S}^{n-1}$ and each $w \in \mathbb{R}^n$ with $w \cdot \omega = 0$ there is a ray coming in from infinity:

$$(5.14) \qquad z(t) = w + t\omega, \ t << 0.$$

This can be continued, possibly in more than one way. The ray will strike the obstacle if w lies in the projection of the obstacle in the direction ω. If the obstacle is contained in the ball $|z| \leq R$ and a continuation of the ray leaves this ball then it will be of the form

$$(5.15) \qquad z(t) = w' + (t - t')\omega', \ \omega' \in \mathbb{S}^{n-1} \text{ for } t >> 0$$

where again $w' \cdot \omega' = 0$. The *ray relation* of the obstacle is the set

$$(5.16) \qquad \Lambda_H = \{(t', \omega', \omega, \tau, \tau w', -\tau w); 0 \neq \tau \in \mathbb{R}\}$$

where there is a ray $z(t)$ of the form (5.14) and (5.15). The value of t' for a particular point in the ray relation is called the *sojourn time* of that ray. [16] It is a measure of the 'additional time' the ray spends in the neighbourhood of the obstacle. The *regular* part of the ray relation is defined by

$$(5.17)$$
$$\Lambda_H^{\text{reg}} = \big\{(t', \omega', \omega, \tau, \tau w', -\tau w); 0 \neq \tau \in \mathbb{R}$$
$$\text{and there is no point of the type (iii) or (v) on the ray}\big\}.$$

The rays defining Λ_H^{reg} are actually those which do not have any point of type (iii) or (iv), since a point of type (iv') cannot occur until one of type (v) has been encountered. The transversality of the reflection at points of type (ii) means that Λ_H^{reg} is a smooth submanifold.[17]

[15] In fact it can be shown, using the dynamical properties of rays, that the set of points which can be reached at a fixed distance along all the continuations of any one ray is always connected. Thus in this case there must be an uncountably infinite set of continuations of the given ray!

[16] This concept was introduced by Guillemin in [29] which was the first systematic application of microlocal analysis to scattering theory. It is still very much worth reading.

[17] I conjecture, having not seen an explicit proof, that Λ_H^{reg} *is precisely the smooth part of* Λ_H.

The fibre of the cotangent bundle of the sphere, $T^*\mathbb{S}^{n-1}$, at the point ω can be identified with the orthocomplement of ω in $\mathbb{R}^n$. Since $T^*(\mathbb{R} \times \mathbb{S}^{n-1} \times \mathbb{S}^{n-1})$ is isomorphic to $T^*\mathbb{R} \times T^*\mathbb{S}^{n-1} \times T^*\mathbb{S}^{n-1}$ the ray relation can be identified as a subset of the cotangent bundle of $\mathbb{R} \times \mathbb{S}^{n-1} \times \mathbb{S}^{n-1}$. The regular part is a Lagrangian submanifold.[18]

Proposition 5.3 [19] *The scattering kernel for an obstacle satisfies*

(5.18) $$\Lambda_H^{\text{reg}} \subset \text{WF}(B_H) \subset \Lambda_H.$$

Given the conjecture in Footnote 17 this means that the regular set Λ_H^{reg} can be recovered from B_H'; indeed it can be recovered from the knowledge of B_H' modulo $\mathcal{C}^\infty$ terms. In any case the closure of Λ_H^{reg} can be recovered from the wavefront set, since the complementary part has smaller Hausdorff dimension.[20] One can therefore ask the natural question as to whether Λ_H^{reg} determines H. That this is not the case is shown by the following example due to Livshits; [21] see also the note of Rauch, [97], and related examples given by Penrose in a somewhat different context.

Example Consider the obstacle in Figure 9. The inner curve between A and B is part of an ellipse, including the half above the major semi-axis, CD; E and F are the foci of this ellipse. The boundary curve is simply tangent to the major semiaxis at E and F and is otherwise just a smooth, say $\mathcal{C}^\infty$, curve, H, which does not meet the major semiaxis CD at any other point. This forms the boundary of the obstacle $\mathcal{O}$. Consider the behaviour of reflected ($\mathcal{C}^\infty$) rays for this obstacle. Any ray passing

---

[18] That is, the symplectic form vanishes when restricted to Λ_H^{reg}.

[19] The first inclusion follows from the fact that B_H' is a Lagrangian distribution, microlocally near Λ_H^{reg}. That is, it can be constructed quite explicitly in terms of oscillatory integrals. The second inclusion follows from a construction as in Section 3.5 which allows B_H to be obtained as the Lax-Phillips transform of a solution to the forcing problem (with Dirichlet boundary condition on $\mathcal{O}$). Results on the propagation of singularities (by Taylor [112] and in [67], [79] and [80], see also [44]) and the calculus of wavefront sets (using the properties of the Radon transform as a Fourier integral operator) then lead to (5.18). Unfortunately I do not know of anywhere this is published in full detail.

[20] That is, the closure of Λ_H^{reg} is necessarily the closure of the subset of $\text{WF}(B_H)$ which is a smooth manifold of dimension $n + 1$. This also suggests some more geometric questions. Since $\text{WF}(B_H)$ is closed, the closure of Λ_H^{reg} is also a subset of Λ_H. Is it the case that $\Lambda_H = \overline{\Lambda_H^{\text{reg}}}$? It seems to me that this should be true. It would imply that

(5.19) $$\text{WF}(B_H) = \Lambda_H.$$

[21] Mikhail Livshits to be precise. Unfortunately this example was not published by him.

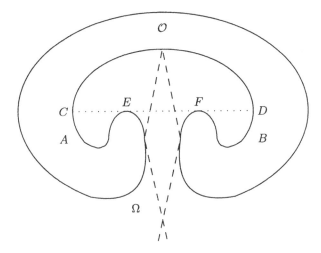

Fig. 9. Two secret rooms.

upward through the segment EF of the major semiaxis is immediately reflected back again, across the same segment.[22] This means that any reflected ray coming in from infinity cannot enter the region of $\mathcal{O}$ lying below FD (or CE.) Thus, from the propagation of singularities result, (5.18), and the stronger result that B_H is determined microlocally by the obstacle in a neighbourhood of the region accessible by rays coming in from infinity, it follows that $\mathrm{WF}(B_H)$, or even B_H modulo $\mathcal{C}^\infty$, cannot detect the structure of the boundary in the open region DF (or CE.) Thus by smoothly modifying the boundary near B one can get two[23] different obstacles with scattering kernels which differ by a $\mathcal{C}^\infty$ [24] function.[25]

---

[22] This is the basic property of the ellipse, that the two lines to the foci from a point on the ellipse make equal angles with the normal to the ellipse at that point. Snell's law then implies that the reflected ray from a point on EF also crosses EF.

[23] Or even an uncountably infinite set of such obstacles.

[24] Quite a lot is now known about the propagation of singularities modulo finer regularity than $\mathcal{C}^\infty$. In this case one would expect that, if the boundary is just short of analytic, in the sense that it is in all Gevrey classes greater than 1 (the obstacle can be constructed in these classes) then the singularities at the level of Gevrey 3 or finer may well detect the difference between two such obstacles. For a discussion of the propagation of Gevrey regularity see [55].

[25] Notice that the 'rooms' bounded by FBD or CAE are 'quiet' (or 'dark') in the sense that no strong waves can reach there from outside. On the other hand the echo in there might be a bit strong.

Despite this counterexample there are several positive results. If the obstacle is convex then Λ_H is the closure of Λ_H^{reg} and the obstacle is easily recovered from it. In fact Majda, [56], see also [57], observed that, assuming H to be smooth and convex, it can be recovered from the backscattering; more precisely it can be recovered from the singularities of $B_H \upharpoonright \mathbb{R} \times \Delta^-(\mathbb{S}^{n-1})$. [26] In fact the same is true if the obstacle is *normally accessible* in the sense that the outward normal from each point of H does not hit the obstacle at any other point.

5.5 Trapped rays

The global behaviour of the reflected geodesics, for instance if there are periodic rays, has a very considerable effect on the scattering properties. One natural condition, the importance of which was emphasized by Lax and Phillips is

Definition 5.2 *An obstacle is said to trap rays, or to have trapped rays, if the sojourn time is unbounded above, i.e. for any T there is an element of the ray relation $(t', \omega', \omega, \tau, \tau w', -\tau w)$ with $t' > T$. If the obstacle docs not trap rays it is said to be 'non-trapping.'*

Thus if the obstacle is non-trapping, the singular support of the scattering kernel is compact. [27] In odd dimensions the Lax-Phillips semigroup can be defined very much as in the case of potential scattering and the argument in Proposition 4.5 carries over to give

Proposition 5.4 *If $n \geq 3$ odd and the obstacle is non-trapping then for any constant N there are only finitely many poles of the resolvent satisfying*

$$(5.20) \qquad \operatorname{Im} \lambda \leq N + N \log(1 + |\operatorname{Re} \lambda|).$$

A question posed by Lax and Phillips concerns the possible converse to Proposition 5.4. They had originally conjectured that an obstacle which trapped rays would necessarily have a sequence of poles with imaginary part tending to zero. This was shown not to be the case by Ikawa,

---

[26] Here $\Delta^-(\mathbb{S}^{n-1})$ is the antidiagonal, where $\omega = -\theta$ in $\mathbb{S}^{n-1} \times \mathbb{S}^{n-1}$. It is also of interest to note that the fact that the obstacle is convex can be determined from the regular part of the ray relation. Namely the convexity of $\mathcal{O}$ is equivalent to the fact that for every $\omega \neq \theta$ there is a unique ray in Λ_H^{reg} linking ω and θ.

[27] This is true for example if the obstacle is *star-shaped* in the sense that there is a fixed point p such that for each $\omega \in \mathbb{S}^{n-1}$ there is exactly one point on H of the form $p + r_\omega \omega$ with $r_\omega > 0$.

see [45], for a simple system of two disjoint convex obstacles. The existence and position of 'pseudo-poles' for this problem had previously been shown by Bardos, Guillot and Ralston, [6], and the distribution of the poles has been more precisely analyzed by Gérard in [26]. Nevertheless various modified forms of the Lax-Phillips conjecture are consistent with present knowledge of the distribution of poles in scattering theory. For example

Modified Lax-Phillips conjecture (A)[28]
For every trapping obstacle there are infinitely many poles in some half-space $\operatorname{Im} \lambda < C$.

Modified Lax-Phillips conjecture (B)
For every trapping obstacle there are infinitely many poles in any region $\operatorname{Im} \lambda \leq N + N \log(1 + |\operatorname{Re} \lambda|)$.[29]

The overall estimates on the distribution of poles are also the same as in the case of potential scattering.

Proposition 5.5 *For $n \geq 3$ odd the counting function for scattering poles, given by (4.23) satisfies (4.25).*[30]

Similarly the trace formula, (4.11), carries over directly

$$(5.21) \qquad \tau_H(t) = \mathcal{F}^{-1} \frac{d}{d\lambda} s_H(\lambda)$$

where $s_H(\lambda)$ is $-i$ times the logarithm of the determinant of the scattering matrix.[31] From this a form of 'Weyl's law' for the large frequency behaviour of the scattering phase can be deduced:

$$(5.22) \qquad s_H(\lambda) = c_n \operatorname{Vol}(\mathcal{O})\lambda^n + O(\lambda^{n-1}) \text{ as } \lambda \to \infty.$$

This is an analogue of (4.18) but there is no full asymptotic expansion in case the obstacle is trapping.[32]

The Poisson formula (4.19) is also valid for obstacle scattering.[33] This has been used to get a variety of estimates on the distribution of scattering poles under various conditions. I will only discuss one such result.[34]

---

[28] By assumption $n \geq 3$ is odd always here.
[29] Of course (A) implies (B)!
[30] This is shown in [71]; this indeed was the first case in which the optimal order bound was proved, even though the obstacle case is the more difficult technically.
[31] So it doesn't matter that this is only determined up to an additive constant.
[32] The result is then rather harder to prove; see [72].
[33] First proved for all $t \neq 0$ in [69]; see Sjöstrand-Zworski [110] for a discussion including 'black-box' perturbations in odd dimensions.
[34] See Zworski's survey [121] for a more complete discussion.

Namely, consider a strictly convex obstacle. Hargé and Lebeau, [34], using the complex scaling method (similar to that used by Sjöstrand and Zworski in [109]) showed that the poles satisfy an estimate [35]

$$(5.23) \qquad \operatorname{Im} \lambda > c|\operatorname{Re} \lambda|^{\frac{1}{3}}, \ c > 0.$$

They even give a geometrically determined value of c for which there are at most a finite number of poles not satisfying (5.23).

---

[35] A similar estimate on any real-analytic obstacle which is non-trapping for C^∞ rays was obtained by Bardos, Lebeau and Rauch [7], see also Popov [95], using the results, on the propagation of singularities with respect to functions of Gevrey class 3, by Lebeau [55].

6

Scattering metrics

In the last three lectures I shall stray beyond the confines of 'traditional,' i.e. Euclidean, scattering theory. I want to show how some of the ideas and methods of Euclidean scattering can be extended to give a detailed description of the generalized eigenfunctions of the Laplacian on certain complete Riemannian manifolds with a regular structure at infinity. I shall discuss, in varying levels of detail, five main classes of such metrics, each being a complete metric on the interior of a compact manifold with boundary. The first two of these, consisting of *asymptotically flat* ('scattering' or 'sc-') metrics and *cylindrical end* ('boundary' or simply 'b-') metrics will be the topics of this, and the penultimate, lecture. In each case I will try to show that these classes of metrics exemplify a general principle. Namely for each of them there are results analogous to the standard differential-geometric theorems on a compact manifold with boundary. These results include a 'Hodge theorem' (description of the null space of the Laplacian on forms) a 'Seeley theorem' (description of the resolvent family as pseudodifferential operators[1]) an 'index theorem' (description of the Fredholm properties of these pseudodifferential operators and[2] a formula of Atiyah-Singer type for the index). In the final lecture I shall try to give an overview of the classification of these metrics.

The class I shall talk about today consists of the scattering metrics. These are direct generalizations of Euclidean spaces. Thus one can expect that the results are rather direct extensions of those discussed above. This is indeed the case, but the possibility of non-trivial topol-

---

[1] Hence the possibility of describing the structure of functions of the operator, e.g. complex powers, see [104] and [105].
[2] At least in some cases!

ogy in the underlying space means that even some things which are simple on Euclidean space manifest interest in the more general setting.

6.1 Manifolds with boundary

The setting for these metrics is an arbitrary compact manifold with boundary, X.[3] The 'trivial case' to bear in mind is $\mathbb{S}^n_+$, obtained by stereographic compactification of $\mathbb{R}^n$. The form of the Euclidean metric on $\mathbb{S}^n_+$ has already been discussed in (1.52). This can be generalized immediately to any compact manifold with boundary. A Riemann metric[4] on X°, the interior of X, is a *scattering metric* if it can be written in the form

$$(6.1) \qquad\qquad g = \frac{dx^2}{x^4} + \frac{h}{x^2}$$

where h is a smooth symmetric 2-cotensor which is positive-definite when restricted to the boundary[5] and $x \in C^\infty(X)$ is a boundary defining function.[6] In fact it is straightforward to check that, once it is known to exist, the form (6.1) of the metric determines $x \in C^\infty(X)$ up to addition

---

[3] A manifold with boundary, by implication always C^∞, is a Hausdorff and paracompact topological space, X, which is locally modelled on a half space, $\mathbb{R}^n_+ = [0,\infty) \times \mathbb{R}^{n-1}$. An open subset $U \subset \mathbb{R}^n_+$ is the intersection of an open subset $U' \subset \mathbb{R}^n$ with $\mathbb{R}^n_+$. A smooth function on U is, by definition, the restriction to U of an element of $C^\infty(U')$ and the space of these functions is denoted $C^\infty(U)$. Similarly a diffeomorphism between open sets of $\mathbb{R}^n_+$ is taken to be the restriction to these sets of a diffeomorphism between open subsets of $\mathbb{R}^n$. Then the C^∞ structure on X is given by a covering of X by open subsets U_j each homeomorphic to an open subset of $\widetilde{U}_j \subset \mathbb{R}^n$ by maps $f_j : U_j \to \widetilde{U}_j$ such that for each j, k with $U_j \cap U_k \neq \emptyset$ the composite map $f_j \circ f_k^{-1} : f_k(U_j \cap U_k) \to f_j(U_j \cap U_k)$ is a diffeomorphism of open subsets of $\mathbb{R}^n_+$.

[4] Meaning of course a positive-definite smooth 2-cotensor on X°.

[5] The tangent and cotangent bundles of a manifold with boundary are defined as in the case of manifolds without boundary. In fact any manifold with boundary can be 'doubled' across the boundary to a manifold without boundary, which I will denote somewhat enigmatically as $2X$. The tangent and cotangent bundles are then naturally the restrictions to X of the bundles over the double. Thus h is supposed to be a smooth section of $T^*X \otimes T^*X$ which is symmetric and smooth up to the boundary in the sense that it is the restriction of a smooth section over the double. The tangent bundle to the boundary is a smooth subbundle of the restriction of TX to the boundary, which I denote $T_{\partial X}X$. Thus the condition that $h_{\partial X} \in C^\infty(X; T^*X \otimes T^*X)$ should restrict to a positive definite form on each space $T_p\partial X \times T_p\partial X \subset T_pX \times T_pX$ for $p \in \partial X$ is meaningful.

[6] Boundary defining functions always exist on a manifold with boundary, that is there is always a function $x \in C^\infty(X)$ which satisfies $x \geq 0$, $x = 0$ on ∂X and $dx \neq 0$ at ∂X. Any two boundary defining functions x and x' are related by $x' = ax$ where $a \in C^\infty(X)$ is everywhere positive.

of a smooth multiple[7] of x^2 and hence fixes $h_{\partial X}$ unambiguously. Thus the boundary ∂X is itself naturally a Riemann manifold[8] when X is given a scattering metric.[9]

6.2 Hodge theorem

One of the most basic theorems in the differential geometry of a compact Riemann manifold Y, without boundary, is the Hodge theorem [10] which identifies the null space of the Laplacian on forms with the cohomology, either in the sense of DeRham or, because of the DeRham theorem, in the sense of singular cohomology with real coefficients. The exterior power bundles[11] $\Lambda^k Y$ carry natural inner products, defined by the metric. The metric also fixes a density, dg, which can be used to integrate functions. This allows the adjoint of the exterior derivative

$$(6.2) \qquad d : C^\infty(Y; \Lambda^k) \longrightarrow C^\infty(Y; \Lambda^{k+1})$$

to be defined as a differential operator of order 1, $\delta = d^* : C^\infty(Y; \Lambda^{k+1}) \longrightarrow C^\infty(Y; \Lambda^k)$ by

$$(6.3)$$
$$\int_Y \langle \delta u, v \rangle dg = \int_Y \langle u, dv \rangle dg \,\, \forall \,\, u \in C^\infty(Y; \Lambda^{k+1}), \,\, v \in C^\infty(Y; \Lambda^k).$$

Theorem 6.1 *(Hodge-Weyl) The null space of $\Delta = (d + \delta)^2 = d\delta + \delta d$ on $C^\infty(Y; \Lambda^k)$ is, for any compact Riemann manifold without boundary, Y, naturally isomorphic to the singular cohomology with real coefficients $H^k(X) = H^k_{\mathrm{sing}}(Y; \mathbb{R})$.*

For a compact manifold with boundary there are extensions of this result to the Laplacian for a metric smooth and non-degenerate up to the boundary. Here I am more interested in the corresponding result for scattering metrics. To state this, recall that there is a long exact sequence in cohomology relating the absolute [12] and relative cohomologies

---

[7] This means that the scattering metric determines a trivialization of the conormal bundle, $N^*\partial X \subset T^*_{\partial X} X$.

[8] Compact and without boundary.

[9] It should also be noted that any compact manifold with boundary can be given a scattering metric. In fact if ∂X is given a Riemann metric then there is a scattering metric on X which induces the given metric on the boundary.

[10] In this generality due to Weyl.

[11] The antisymmetric parts of the tensor powers of T^*Y.

[12] In deRham form the absolute cohomology is the cohomology of d acting on $C^\infty(X; \Lambda^*)$, forms smooth up to the boundary; the relative cohomology is the

of X with the cohomology of the boundary:

(6.4)
$$\cdots \longrightarrow H^{k-1}(\partial X) \longrightarrow H^k_{\mathrm{rel}}(X) \longrightarrow H^k_{\mathrm{abs}}(X) \longrightarrow H^k(\partial X) \longrightarrow \cdots .$$

Here the coefficients are real throughout. The central map is, in terms of DeRham cohomology, derived from an inclusion map.

Theorem 6.2 *The null space of the Laplacian $\Delta = d\delta + \delta d$, for a scattering metric, acting on square-integrable k-forms is naturally isomorphic to the image, in (6.4), of the relative in the absolute cohomology in dimension k.*

Such a result depends on the construction of a good parametrix.[13] I shall briefly discuss this in the context of the properties of the resolvent family in the physical region.

6.3 Pseudodifferential operators

The treatment of the resolvent family for $\Delta + V$ in Lecture 2, and to a lesser extent in the subsequent results on Euclidean scattering, depends heavily on the representation (2.2) which expresses it in terms of the free resolvent. To get information on the free resolvent, as in Lecture 1, its kernel can be computed explicitly, or, as I have done, its representation in terms of the Fourier transform can be used. For the resolvent family for the Laplacian of a scattering metric such direct methods are not available.[14] Instead I shall describe an algebra of 'scattering' pseudodifferential operators on X which are tailored to contain the resolvent family. That the resolvent does takes values in this space can be shown constructively.

Let me first describe this algebra of pseudodifferential operators on $\mathbb{R}^n$. This will allow the definition in the general case to be made by

---

cohomology of d acting on $\dot{C}^\infty(X; \Lambda^*)$, forms vanishing to infinite order at the boundary, or equivalently $C^\infty_c(X^\circ, \Lambda^*)$, forms with compact support in the interior.

[13] That is, an approximate inverse, where the 'good' refers to the error being appropriately compact.

[14] Although there is certainly a possible replacement. Namely one could (i.e. I haven't) consider the special 'model scattering metrics' where X is replaced by $[0, \infty) \times Y$, with h in (6.1) taken to be independent of x and to have no factors of dx. Then 'separation of variables' allows the Laplacian for the full metric to be written as a function of the Laplacian for h on Y. Indeed this can be done in terms of Bessel functions. Now one can analyze the behaviour of this function 'directly' and (probably) use perturbation theory to examine the original problem. I prefer to proceed to construct the kernel of the resolvent more explicitly as outlined here.

localization, since X is everywhere locally diffeomorphic to $\mathbb{S}^n_+$. Pseudo-differential operators were originally defined as operators on functions on $\mathbb{R}^n$ through the Fourier transform[15]

(6.5)
$$Au = (2\pi)^{-n} \int_{\mathbb{R}^n} e^{iz\cdot\zeta} a(z,\zeta)\widehat{u}(\zeta)d\zeta,$$
$$\widehat{u}(\zeta) = \int_{\mathbb{R}^n} e^{-i\zeta\cdot z}u(z)dz.$$

The class of operators obtained depends on the conditions imposed on the 'amplitude' (or full symbol) a. The class arising here was considered explicitly by Shubin [106] and Parenti [88]. It corresponds to functions a which are symbols of 'product type' in z and ζ with full asymptotic expansions at infinity.[16] This can be easily expressed in terms of the stereographic compactification. Thus consider separate stereographic compactification in each of these variables

(6.7)
$$\mathrm{SP}^2 : \mathbb{R}^n \times \mathbb{R}^n \longrightarrow \mathbb{S}^n_+ \times \mathbb{S}^n_+.$$

The image space is a manifold with corners, and hence $\mathcal{C}^\infty(\mathbb{S}^n_+ \times \mathbb{S}^n_+)$ is naturally defined.[17] Let ρ_∂ and ρ_∞ be respective boundary defining functions for the two boundary hypersurfaces $\mathbb{S}^{n-1}\times\mathbb{S}^n_+$ and $\mathbb{S}^n_+\times\mathbb{S}^{n-1}$.[18] Then

Definition 6.3 *For any $m, k \in \mathbb{R}$, the space $\Psi^{m,k}_{\mathrm{sc}}(\mathbb{S}^n_+)$ consists of the operators of the form (6.5) with amplitude*

$$a \in (\mathrm{SP}^2)^* \rho_\partial^k \rho_\infty^{-m} \mathcal{C}^\infty(\mathbb{S}^n_+ \times \mathbb{S}^n_+).$$

The residual operators, those in

(6.8)
$$\Psi^{-\infty,\infty}_{\mathrm{sc}}(\mathbb{S}^n_+) = \bigcap_{k,m} \Psi^{m,k}_{\mathrm{sc}}(\mathbb{S}^n_+),$$

---

[15] See the original papers by Kohn and Nirenberg [49] and Hörmander [38] as opposed to the older, more restricted, definitions in terms of singular integral operators.

[16] The properties of this 'calculus' on $\mathbb{R}^n$ follow from the general 'Weyl calculus' of Hörmander, see [41] or [44]. It corresponds to the slowly varying metric

(6.6)
$$\frac{|dz|^2}{1+|z|^2} + \frac{|d\zeta|^2}{1+|\zeta|^2}.$$

[17] Precisely, $a \in \mathcal{C}^\infty(\mathbb{S}^n_+ \times \mathbb{S}^n_+)$ if and only if there exists a function $u' \in \mathcal{C}^\infty(\mathbb{S}^n \times \mathbb{S}^n)$ such that $u = u' \upharpoonright (\mathbb{S}^n_+ \times \mathbb{S}^n_+)$.

[18] For instance take $\rho_\partial = (1+|z|^2)^{-\frac{1}{2}}$ and $\rho_\infty = (1+|\zeta|^2)^{-\frac{1}{2}}$.

are precisely those with Schwartz kernels in [19]

$$\mathcal{S}(\mathbb{R}^{2n}) = (\mathrm{SP}^2)^* \left(\dot{\mathcal{C}}^\infty(\mathbb{S}_+^n \times \mathbb{S}_+^n) \right).$$

Furthermore it can be seen that $\Psi_{\mathrm{sc}}^{k,m}(\mathbb{S}_+^n)$ is invariant under diffeomorphisms of $\mathbb{S}_+^n$. This allows a corresponding space of pseudodifferential operators to be defined on any compact manifold with boundary by localization; a similar definition has also been considered by Erkip and Schrohe, [20].

Definition 6.4 [20] *For any compact manifold with boundary, $\Psi_{\mathrm{sc}}^{m,k}(X)$ consists of those continuous linear operators*

(6.9) $$A : \dot{\mathcal{C}}^\infty(X) \longrightarrow \dot{\mathcal{C}}^\infty(X)$$

with the following two properties. First, if ϕ_1, $\phi_2 \in \mathcal{C}^\infty(X)$ have disjoint supports then[21]

(6.10)

$$A'u = \phi_1 A \phi_2 u = \int_X A(p,p')u(p')dg(p') \ \text{ with } A' \in \dot{\mathcal{C}}^\infty(X \times X).$$

Secondly, if $F : U \longrightarrow U'$ is a diffeomorphism between open sets $U \subset X$ and $U' \subset \mathbb{S}_+^n$, $\phi \in \mathcal{C}^\infty(X)$ has support in U and $\phi' \in \mathcal{C}^\infty(\mathbb{S}_+^n)$ has support in U' then

(6.11) $$\dot{\mathcal{C}}^\infty(\mathbb{S}_+^n) \ni v \longmapsto A''v = (F^{-1})^* \left(\phi A(F^*(\phi' v)) \right) \in \dot{\mathcal{C}}^\infty(\mathbb{S}_+^n)$$

is an element of $\Psi_{\mathrm{sc}}^{m,k}(\mathbb{S}_+^n)$.

---

[19] This space can in turn can be identified with the subspace of $\mathcal{C}^\infty(\mathbb{s}^n \times \mathbb{s}^n)$ consisting of the functions which vanish identically outside $\mathbb{s}_+^n \times \mathbb{s}_+^n$.

[20] There is a different, more global and 'intrinsic' definition which I prefer. This sort of definition can be used for all the calculi of pseudodifferential operators which I briefly discuss in the remaining lectures. In barest outline the definition is based on 'blow-up.' The kernels of pseudodifferential operators in the ordinary sense are conormal distributions (in fact distributional densities) at the diagonal of the product X^2. This manifold can be replaced by a blown up, or stretched, version denoted X_f^2 where 'f' refers to the structure in question; this is a compact manifold with corners. In this case f=sc. The construction of X_f^2, in each circumstance, is based on the properties of the Lie algebra of vector fields $\mathcal{V}_f(X)$ underlying the problem. The f-pseudodifferential operators are then defined by conormal distributions with respect to a 'lifted diagonal' in X_f^2. Examples of this construction can be found in [62], [76] and [60]; the general theory should be contained in [66].

[21] Here I have used the Riemann density dg of some scattering metric. The space $\dot{\mathcal{C}}^\infty(X \times X) \subset \mathcal{C}^\infty((2X) \times (2X))$ consists of the functions vanishing identically outside $X \times X$.

These spaces of operators are invariant under diffeomorphism of X and form a bifiltered algebra

$$(6.12) \qquad \Psi_{\mathrm{sc}}^{m_1,k_1}(X) \circ \Psi_{\mathrm{sc}}^{m_2,k_2}(X) \subset \Psi_{\mathrm{sc}}^{m_1+m_2,k_1+k_2}(X).$$

The definition is designed so that

Theorem 6.3 *For any scattering metric (6.1) on a compact manifold with boundary the resolvent family of the Laplacian is such that*

$$(6.13)$$
$$\{\lambda \in \mathbb{C}; \operatorname{Im} \lambda < 0\} \ni \lambda \longmapsto R(\lambda) = (\Delta - \lambda^2)^{-1} \in \Psi_{\mathrm{sc}}^{-2,0}(X)$$

is a holomorphic family.

This is the direct extension to this 'category' of Seeley's theorem, which says the same thing for compact manifolds without boundary.

6.4 Symbol calculus

A result such as Theorem 6.3 can be proved using the symbolic and related properties of the algebra $\Psi_{\mathrm{sc}}^{*,*}(X)$. To describe some of these let me first introduce the natural replacement for $\mathbb{S}_+^n \times \mathbb{S}_+^n$, in the case of a general compact manifold with boundary X, as the carrier of the symbols in (6.5) and (6.7).

To do so consider the elementary properties of a scattering metric, (6.1). Let $\mathcal{V}(X) = \mathcal{C}^\infty(X; TX)$ be the space of smooth vector fields on X. These are the restrictions to X of smooth vector fields on $2X$, its double to a manifold without boundary. As such they have no special properties with respect to the boundary. Let $\mathcal{V}_{\mathrm{b}}(X) \subset \mathcal{V}(X)$ be the subspace of vector fields which are tangent to the boundary.[22] Now consider the elements of $\mathcal{V}(X)$ with bounded length with respect to a scattering metric. Then[23]

$$(6.14)$$
$$\mathcal{V}_{\mathrm{sc}}(X) = \{V \in \mathcal{V}(X); |V|_g \text{ is uniformly bounded on } X^\circ\} = x\mathcal{V}_{\mathrm{b}}(X).$$

This space of vector fields, $\mathcal{V}_{\mathrm{sc}}(X)$, is a Lie algebra [24] under the commutator bracket. It is also the case that there is a natural vector bundle

---

[22] If $x \in \mathcal{C}^\infty(X)$ is a boundary defining function and $V \in \mathcal{V}(X)$ then $V \in \mathcal{V}_{\mathrm{b}}(X)$ if and only if $Vx \in x\mathcal{C}^\infty(X)$, i.e. Vx vanishes at ∂X.

[23] The first inequality here is the definition.

[24] This follows directly from the second equality in (6.14), since if $V = xV'$ and $W = xW'$ are elements of $\mathcal{V}_{\mathrm{sc}}(X)$ so is

$$(6.15) \qquad [V, W] = [xV', xW'] = x^2[V', W'] + x(V'x)W' - x(W'x)V'$$

over X, smooth up to the boundary, which I shall denote $^{\mathrm{sc}}TX$, with the defining property[25]

$$(6.18) \qquad \mathcal{V}_{\mathrm{sc}}(X) = \mathcal{C}^\infty(X; {}^{\mathrm{sc}}TX).$$

The bundle $^{\mathrm{sc}}TX$ is independent of which scattering metric is used to define it.[26] It is therefore more fundamental than the notion of a scattering metric. I call it the *scattering tangent bundle.* [27] Let $^{\mathrm{sc}}T^*X$ be the dual bundle to $^{\mathrm{sc}}TX$; it is the 'scattering cotangent bundle'. The fibre of $^{\mathrm{sc}}T^*X$ is a vector space of dimension $n = \dim X$ over each point. Let $^{\mathrm{sc}}\bar{T}^*X$ be the compact manifold with corners obtained by replacing each fibre by its stereographic compactification to a half sphere. For $X = \mathbb{S}^n_+$ it is certainly the case that $^{\mathrm{sc}}\bar{T}^*\mathbb{S}^n_+ = \mathbb{S}^n_+ \times \mathbb{S}^n_+$. In the general case $^{\mathrm{sc}}\bar{T}^*X$ is a replacement for this space. Let $C_{\mathrm{sc}}X = \partial(^{\mathrm{sc}}\bar{T}^*X)$ be the boundary of this compact manifold with corners; it is not a manifold with boundary but is rather the union of two[28] manifolds with boundary, with the boundaries of these two manifolds naturally identified. As notation set

$$(6.19) \qquad \begin{aligned} \mathcal{C}^\infty(C_{\mathrm{sc}}X) = \\ \left\{ u : C_{\mathrm{sc}}X \longrightarrow \mathbb{C}; \ \exists\, \tilde{u} \in \mathcal{C}^\infty(^{\mathrm{sc}}\bar{T}^*X) \text{ with } u = \tilde{u} \restriction C_{\mathrm{sc}}X \right\}. \end{aligned}$$

This space is defined precisely so that

---

since $\mathcal{V}_{\mathrm{b}}(X)$ itself is a Lie algebra, see Lecture 7.

[25] Really $^{\mathrm{sc}}TX$ is a vector bundle which comes equipped with a vector bundle map $\iota_{\mathrm{sc}} : {}^{\mathrm{sc}}TX \longrightarrow TX$ which is an isomorphism over X° and such that for $V \in \mathcal{V}(X) = \mathcal{C}^\infty(X;TX)$, $V \in \mathcal{V}_{\mathrm{sc}}(X)$ is equivalent to V lifting to a section of $^{\mathrm{sc}}TX$; this is how to interpret (6.18). The definition of the fibre, $^{\mathrm{sc}}T_pX$, for any $p \in X$ is through the ideal $\mathcal{I}_p(X) \subset \mathcal{C}^\infty(X)$ of functions vanishing at p. Then

$$(6.16) \qquad {}^{\mathrm{sc}}T_pX = \mathcal{V}_{\mathrm{sc}}(X)\big/\mathcal{I}_p(X) \cdot \mathcal{V}_{\mathrm{sc}}(X)$$

where the subspace $\mathcal{I}_p(X) \cdot \mathcal{V}_{\mathrm{sc}}(X)$ is the finite linear span of products. If x, $y_1, \ldots, y_{n-1}$ are local coordinates near a boundary point, with x a local boundary defining function then the vector fields

$$(6.17) \qquad x^2 \frac{\partial}{\partial x}, x\frac{\partial}{\partial y_1}, \ldots, x\frac{\partial}{\partial y_{n-1}}$$

form a local, smooth, basis for $^{\mathrm{sc}}TX$.

[26] Since all such metrics give the same set of vector fields of bounded length.

[27] Notice that, essentially by definition, a scattering metric defines a non-degenerate fibre metric on $^{\mathrm{sc}}TX$. The converse is not quite true, because of the special splitting of normal and tangential parts of the metric implicitly in (6.1).

[28] Each of which is itself possibly disconnected.

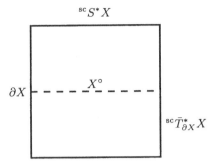

Fig. 10. The compactified scattering cotangent bundle.

Lemma 6.1 *There is a natural map*[29]

$$(6.20) \qquad \sigma_{\mathrm{sc},0,0} : \Psi_{\mathrm{sc}}^{0,0}(X) \longrightarrow \mathcal{C}^{\infty}(C_{\mathrm{sc}}X)$$

which gives a short exact sequence

$$(6.21) \qquad 0 \longrightarrow \Psi_{\mathrm{sc}}^{-1,1}(X) \hookrightarrow \Psi_{\mathrm{sc}}^{0,0}(X) \overset{\sigma_{\mathrm{sc},0,0}}{\longrightarrow} \mathcal{C}^{\infty}(C_{\mathrm{sc}}X) \longrightarrow 0.$$

There are various direct extensions of this result. First, if E is a $\mathcal{C}^{\infty}$ vector bundle over X then the spaces $\Psi_{\mathrm{sc}}^{m,k}(X;E)$ are easily defined.[30] Similarly the symbol map and symbol sequence extend to other orders. Thus, for any $m, k \in \mathbb{R}$, the space $\rho_{\partial}^{k}\rho_{\infty}^{-m}\mathcal{C}^{\infty}({}^{\mathrm{sc}}\bar{T}^{*}X)$ can be interpreted as the space of all smooth sections of a line bundle over ${}^{\mathrm{sc}}\bar{T}^{*}X$, which I shall denote $Q_{m,k}$. For general order and vector bundle the map (6.20) and sequence (6.21) become[31]

$$(6.22)$$
$$0 \longrightarrow \Psi_{\mathrm{sc}}^{m-1,k+1}(X;E) \hookrightarrow \Psi_{\mathrm{sc}}^{m,k}(X;E)$$
$$\overset{\sigma_{\mathrm{sc},m,k}}{\longrightarrow} \mathcal{C}^{\infty}(C_{\mathrm{sc}}X; \pi^{*}\hom(E) \otimes Q_{m,k}) \longrightarrow 0.$$

These symbol maps are multiplicative under composition.

---

[29] This is the joint symbol map; it really consists of two pieces, one being the symbol in the ordinary sense and the other coming from the behaviour of the operator at the boundary, i.e. 'spatial infinity' in $\mathbb{R}^n$.

[30] Just use the local definition as in Definition 6.4 with open sets over which E is trivial so that the corresponding operator on $\mathbb{S}_+^n$ is a matrix of operators in the sense of Definition 6.3.

[31] Here $\pi : C_{\mathrm{sc}}X \longrightarrow X$ is the projection.

This is quite analogous to the case of a compact manifold with boundary, except for the 'extra' part of the symbol coming from ∂X. Various other of the standard constructions extend easily. For example, let $H^k_{sc}(\mathbb{S}^n_+) = \mathrm{SP}^* \left(H^k(\mathbb{R}^n) \right)$ be the image under stereographic compactification of the standard Sobolev spaces[32]on $\mathbb{R}^n$. This can be transferred, by localization, to sections of any vector bundle over a general compact manifold with boundary so defining the spaces $H^k_{sc}(X; E)$. More generally, set $H^{k,l}_{sc}(X; E) = x^l H^k_{sc}(X; E)$ for $l \in \mathbb{R}$ and x a boundary defining function. Then the scattering pseudodifferential operators are always bounded as maps

$$
\text{(6.23)} \qquad \Psi^{m,k}_{sc}(X; E) \ni A : H^{M,K}_{sc}(X; E) \longrightarrow H^{M-m,K+k}_{sc}(X; E)
$$
$$
\forall\, m, k, M, K \in \mathbb{R}.
$$

6.5 Index theorem

An element, $A \in \Psi^{m,k}_{sc}(X; E)$, of the ring of scattering pseudodifferential operators acting on sections of a vector bundle E, is Fredholm as an operator (6.23) if, and only if, it is 'totally' elliptic in the sense that its symbol, $\sigma_{sc,m,k}(A)$ is an isomorphism at each point of $C_{sc}X$.[33] As a compact manifold with corners $^{sc}\bar{T}^*X$ can be retracted (or smoothed) to a compact manifold with boundary $Z'(X)$. The bundle isomorphism $\sigma_{sc,m,k}(A)$ defines an isomorphism of a bundle over the boundary of $Z'(X)$, up to homotopy. Using the clutching construction[34] this data in turn defines a virtual vector bundle, over the double, $Z(X)$, of $Z'(X)$ and hence an element $\mathrm{AS}(A) \in K^0(Z(X))$ of the K space of $Z(X)$. In fact $\mathrm{AS}(A)$ arises from a class with support in Z'. This relative class on T^*X is the Atiyah-Singer class. The full index class is $\mathrm{AS}(A)\,\mathrm{Td}(T^*X)$ where $\mathrm{Td}(T^*X)$ is the Todd class of T^*X, thought of as an absolute class on Z.

---

[32] Take $k \geq 0$ for simplicity so that the elements are functions. In fact if $k < 0$ then the elements of $H^k_{sc}(\mathbb{S}^n_+)$ can be defined in the same way, as 'extendible distributions.'

[33] Notice that the Laplacian itself is not 'totally' elliptic since its symbol, just the square of the metric length function on $^{sc}T^*X$, vanishes at the zero section, which meets $C_{sc}X$ over the boundary; on the other hand $\Delta + 1$ is totally elliptic.

[34] See [4] or [5].

Theorem 6.4 [35] *The index of a (totally) elliptic element* $A \in \Psi_{\mathrm{sc}}^{m,k}(X;E)$
is the integer

$$(6.24) \qquad \mathrm{ind}(A) = \langle \mathrm{AS}(A)\,\mathrm{Td}(T^*X), [T^*X] \rangle$$

obtained by pairing the index class of the symbol with the fundamental class.

In case the boundary of the manifold is empty this is just the Atiyah-Singer theorem. This result should be compared with the Atiyah-Patodi-Singer index theorem discussed in Lecture 7.

Given the properties of the algebra of operators $\Psi_{\mathrm{sc}}^{*,*}(X;E)$ the proof of (6.24) is quite straightforward. There are invertible operators of any multiorder m, k and using these it suffices to consider the result for $\Psi_{\mathrm{sc}}^{0,0}(X;E)$. Any elliptic operator in $\Psi_{\mathrm{sc}}^{0,0}(X;E)$ can be deformed through such elliptic operators to an element of $\Psi_{\mathrm{sc}}^{0,0}(X;E)$ which is a bundle isomorphism, i.e. multiplication by a matrix, everywhere near the boundary. Then formula (6.24) *is* the Atiyah-Singer formula and its validity in general follows from its homotopy invariance.

6.6 Limiting absorption principle

Now, consider something closer to the 'real' questions of spectral and scattering theory for the Laplacian.[36] As a weak converse to Theorem 6.3[37] let me first note that the spectrum of Δ consists precisely of the half line $[0, \infty)$. Furthermore there is no embedded spectrum, i.e.

---

[35] This result includes several known index theorems. In particular the index theorem of Callias [14] is the special case when $X = \mathbb{S}_+^n$ is, for n odd, obtained by stereographic compactification and A is the Dirac operator on $\mathbb{R}^n$ with an additional term of order zero ensuring its ellipticity in this scattering sense. Callias theorem was extended by Anghel [2] and further extended by Anghel [3] and Råde [96]. Råde's theorem is stated for Dirac operators on general odd-dimensional complete Riemannian manifolds but the invertibility conditions imposed outside a hypersurface allow this result to be reduced, by a simple homotopy, to a special case of Theorem 6.4. These results for Dirac operators therefore bear much the same relationship to Theorem 6.4 as the Atiyah-Singer index theorem for Dirac operators on a compact manifold (with Hermitian Clifford module) bears to the general Atiyah-Singer index theorem for elliptic pseudodifferential operators.

[36] For simplicity acting on functions but these results all have relatively direct extensions to the Laplacian acting on forms. The results of this, and the next, section come from [74].

[37] In the sense that $R(\lambda)$ cannot extend to a larger set in λ as a holomorphic function with values in the bounded operators on $L^2(X; dg) = H_{\mathrm{sc}}^0(X)$.

there are no L^2 eigenfunctions corresponding to $\lambda > 0$.[38] There is in this setting a direct extension of the limiting absorption principle.

Proposition 6.1 *For a scattering metric, as $\lambda' \in \mathcal{P}$ tends to a limiting value $0 \neq \lambda \in \mathbb{R}$ the function $R(\lambda')f$ for any $f \in \dot{C}^\infty(X)$, converges uniformly on compact sets and defines a limiting operator*

$$(6.25) \qquad R(\lambda) : \dot{C}^\infty(X) \longrightarrow C^\infty(X^\circ), \; 0 \neq \lambda \in \mathbb{R}.$$

Moreover,

$$(6.26) \qquad R(\lambda)f = e^{-i\lambda/x} x^{\frac{1}{2}(n-1)} a(x, \lambda) \; \text{with } a \in C^\infty(X).$$

This is the direct analogue of Lemma 2.4 for a potential perturbation of the Euclidean Laplacian.

6.7 Generalized eigenfunctions

The property (6.26) means that one can quite easily find a 'functional parametrization' of the generalized eigenspace of the Laplacian for a scattering metric.

Proposition 6.2 [39] *If $0 \neq \lambda \in \mathbb{R}$ the space*

$$(6.28) \qquad \text{null}(\Delta - \lambda^2) = \left\{ u \in H_{sc}^{\infty, \infty}(X); (\Delta - \lambda^2)u = 0 \right\}$$

is contained in $H_{sc}^{\infty, -\frac{1}{2}+\epsilon}(X)$ for any $\epsilon < 0$ and meets $H_{sc}^{\infty, -\frac{1}{2}+\epsilon}(X)$ in $\{0\}$ for any $\epsilon > 0$. For each $g \in C^\infty(\partial X)$ there is a unique element of $\text{null}(\Delta - \lambda^2)$ of the form[40]

$$(6.29)$$
$$u = e^{i\lambda/x} x^{\frac{1}{2}(n-1)} g + e^{-i\lambda/x} x^{\frac{1}{2}(n-1)} g_- + u', \; g_- \in C^\infty(\partial X), \; u' \in H_{sc}^{0,0}(X)$$

---

[38] This is also true of the action on forms; the only difference is that in the case of functions there is also no L^2 null space, as shown by Theorem 6.2, whereas for the action on forms the same result shows that there may indeed be L^2 null space.

[39] To construct these generalized eigenfunctions it is enough to show that there is a 'formal solution'

$$(6.27) \qquad \qquad v = e^{i\lambda/x} x^{\frac{1}{2}(n-1)} h, \; h \in C^\infty(X), \; h \restriction \partial X = g$$
$$(\Delta - \lambda^2)v \in \dot{C}^\infty(X)$$

and then to use Proposition 6.1 to solve away the error term.

[40] Both g and g_- should be extended to smooth functions in the interior to make sense of (6.29), but the result is independent of how this is done. These solutions necessarily have full asymptotic expansions at the boundary.

and these solutions are dense[41] *in the topology of* $H_{\mathrm{sc}}^{\infty,-\frac{1}{2}+\epsilon}(X)$ *for any* $\epsilon < 0$.

6.8 Scattering matrix

As in Euclidean scattering, the uniqueness of the solution (6.29) for given $g \in C^\infty(\partial X)$ means that there is a *scattering matrix*

$$(6.30) \qquad A(\lambda) : C^\infty(\partial X) \ni g \longmapsto g_- \in C^\infty(\partial X).$$

It is easily seen to be unitary with respect to the L^2 structure on the boundary. However, the construction of the solutions (6.29) in this general case is much less direct than in the various cases of Euclidean scattering considered earlier. Therefore the structure of the scattering matrix is not immediately obvious.

Theorem 6.5 [42] *For any* $0 \neq \lambda \in \mathbb{R}$ *the scattering matrix for a scattering metric is a Fourier integral operator, of order 0, associated to geodesic flow on the boundary at distance* π.

To understand where the geodesic flow on the boundary of X comes from (and indeed to prove this result) consider the behaviour of geodesics in the interior of X. By examining the Hamilton equations for geodesics one can show that for a given compact subset $K \subset X^\circ$ there are many geodesics which do not meet K. If K is chosen large enough (a sort of 'compact core') then every geodesic for the given scattering metric on X which does not meet K has a limit point on ∂X in both directions; if x is a given boundary defining function then this is the case for the compact set $\{x \geq \epsilon\}$ if $\epsilon > 0$ is small enough. For a given point on ∂X there is an infinite family of geodesics having this as limit point in a given direction. Furthermore, if one considers a sequence χ_n of geodesics with a fixed limit point at one end,

$$(6.31) \qquad \lim_{t \to -\infty} \chi_n(t) = p \in \partial X \ \forall \ n$$

---

[41] In fact if $g \in C^{-\infty}(\partial X)$ is an arbitrary distribution, which is extended to an element of $C^{-\infty}(X)$ to be constant in the normal direction for some product decomposition near the boundary then there is a unique solution of the form (6.29), except that the 'error term' has should be taken to be in $H_{\mathrm{sc}}^{\infty,-\frac{1}{2}+\epsilon}(X)$ for $0 < \epsilon < 1$. The linear space of these solutions is exactly the null space in (6.28). Thus the null space is identified with $C^{-\infty}(\partial X)$.

[42] This is a result of joint work with Maciej Zworski, [82].

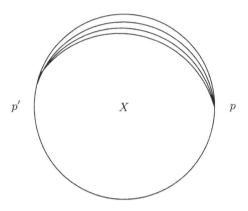

Fig. 11. Geodesic of a scattering metric.

and supposes that for some defining function x the supremum of x on χ_n tends to zero with n[43]

$$(6.32) \qquad \sup_{t\in(-\infty,\infty)} x(\chi_n(t)) \to 0 \text{ as } n \to \infty$$

then any limit point of the other end points of the geodesics, $p_{n_k} \to p'$, $p_n = \lim_{t\to\infty} \chi_n(t)$, where $n_k \to \infty$ as $k \to \infty$, is at distance π from p in the sense that there is a geodesic of length π on ∂X with initial point p and p' as endpoint. The constant π arises from the intrinsic scaling factor in (6.1). In fact much more is true; any subsequence of χ_n has a subsequence which converges (locally uniformly) to a geodesic of length π in the boundary and conversely every boundary geodesic of length π arises in this way. As I shall illustrate in the last lecture with some more examples, at least in the regular cases described in these lectures the singularities of the kernel of the scattering matrix always arise from geodesics on X which are uniformly close to the boundary.

---

[43] So the geodesics are uniformly close to the boundary as $n \to \infty$.

6.9 Long-range potentials

Most of what I have described here for the Laplacian carries over to $\Delta + V$, provided $V \in \mathcal{C}^{\infty}(X)$ vanishes to second order on ∂X.[44] If $V = xV'$ where x is the defining function in (6.1) and $V' \in \mathcal{C}^{\infty}(X)$ is *constant* on ∂X then again the results on the continuous spectrum carry over, except that now there can be discrete spectrum in $(-\infty, 0)$ which can even accumulate at 0. The constructive methods used to prove Theorem 6.5 are closely related to those developed to discuss long-range potentials in Euclidean space; see for example the work of S. Agmon, [1], and Agmon and Hörmander. It is possible to combine these methods with those of [74] and [82] to analyze $\Delta + V$ when V is long-range in this sense.

6.10 Other theorems?

The basic results of scattering on Euclidean spaces, including the limiting absorption principle and parametrization of the continuous spectrum, have been shown to extend to this wider setting. It is therefore natural to ask whether some of the other known results also have such generalizations.

First, there is the question of the possible analytic continuation of the resolvent. It is quite reasonable to expect at least a local continuation through the spectrum, away from 0, provided the metric is supposed to have a product decomposition near the boundary.[45] In general one should expect 0 to be an essential singularity for any such analytic continuation.

In cases of perturbations of the Euclidean metric on $\mathbb{R}^n$ which decay sufficiently rapidly at infinity, analogues of the trace formula are known. See the work of Phillips ([90], [91]) and especially that of Robert ([101], [100]).

---

[44] This is a 'short-range' condition.
[45] See Footnote 14.

7
Cylindrical ends

The second class of complete Riemann metrics I shall consider are those on the interior of a compact manifold with boundary giving a neighbour-hood of the boundary the appearance of a 'cylindrical end'. In fact the class with which I shall work have asymptotically cylindrical ends; I call them 'exact b-metrics'. If X is a compact manifold with boundary[1] and $x \in C^\infty(X)$ is a boundary defining function then a metric on the interior of X is an exact b-metric if, near the boundary, it takes the form

$$(7.1) \qquad g = \frac{dx^2}{x^2} + g' \text{ with } h = g' \restriction \partial X \text{ a metric on } \partial X.$$

Here g' is supposed to be a smooth 2-cotensor. In fact taking g' to be a metric on X in the usual sense, i.e. smooth and non-degenerate up to the boundary, and using (7.1) as a global definition on X shows that any compact manifold with boundary admits such an exact b-metric. The terms 'exact' and the 'b' (which just stands for boundary) are explained below. The traditional approach to metrics of this type is to take a product decomposition of the manifold near its boundary with x as the normal variable and then use (7.1) as the definition of the metric, with $g' = h$ the pull-back of a metric on ∂X. This is a metric of 'product type,' near the boundary. These also always exist and for many questions (e.g. those which are homotopy invariant) it is enough to treat this product-type case. Notice that introducing $t = -\log x$ as variable in place of x the boundary is moved to $t = \infty$ and the metric, in the product case, takes the form

$$(7.2) \qquad\qquad\qquad g = dt^2 + h.$$

---

[1] To simplify the notation I shall assume that the boundary of X is connected, the modifications to handle the general case are very minor, see Footnote 24.

This t variable is natural from the viewpoint of geodesic length and then the perturbation in the more general case (7.1) is seen to be exponentially decreasing as $t \to \infty$. Note that (7.2) shows the manifold equipped with an exact b-metric to look, near the boundary, like the product of a 1-dimensional Euclidean space (i.e. the half line) and a compact manifold with boundary. Thus the term 'cylindrical end.' The analytic properties of the Laplacian (even in the more general case (7.1)) strongly reflect this approximate product decomposition.

7.1 b-geometry

The reason I prefer to think of these metrics in the form (7.1), rather than the form (7.2), is their connection to the intrinsic geometry[2] of the compact manifold with boundary, X. These metrics are, in a certain sense, the most natural complete metrics on the interior of X. I have already briefly described the Lie algebra $\mathcal{V}_b(X)$ of all those smooth vector fields on X which are tangent to the boundary. In local coordinates near the boundary $x, y_1, \ldots, y_p$ with $x \geq 0$ a boundary defining function, $\mathcal{V}_b(X)$ is locally spanned by

$$(7.3) \qquad x\frac{\partial}{\partial x}, \frac{\partial}{\partial y_1}, \ldots, \frac{\partial}{\partial y_p}.$$

From (7.1) it is clear that $\mathcal{V}_b(X)$ is also the space of smooth vector fields on X with bounded length with respect to any exact b-metric. The fact that $\mathcal{V}_b(X)$ has a local basis, (7.3),[3] means that it is in fact the space of (all) smooth sections of a vector bundle, ^bTX, on X.[4] As in the scattering case described in Lecture 6 there is a natural vector bundle map $^bTX \longrightarrow TX$ which is an isomorphism over the interior of X but in this case drops rank by one over the boundary.[5] That the length,

---

[2] Hence I think of them as exact b-metrics on manifolds with boundary rather than as manifolds with asymptotically cylindrical ends, except that I have not explained the 'exact' part (see Footnote 6). You are 'supposed' to think of b-metrics, b-geometry, etc. more or less as a category.

[3] Together with the fact that the coefficients with respect to these local bases are related by smooth matrices under change of coordinates system of this type.

[4] If $p \in X$ (including possibly $p \in \partial X$) and $\mathcal{I}(p) \subset C^\infty(X)$ is the ideal of functions vanishing at p let $\mathcal{I}(p) \cdot \mathcal{V}_b(X)$ be the subspace of $\mathcal{V}_b(X)$ consisting of finite sums of products fV, $f \in \mathcal{I}(p)$ and $V \in \mathcal{V}_b(X)$. The fibre of this vector bundle at p can be defined abstractly as

$$(7.4) \qquad {}^bT_pX = \mathcal{V}_b(X)/\mathcal{I}(p) \cdot \mathcal{V}_b(X).$$

[5] Namely $x\partial/\partial x$ spans the null space at each boundary point.

with respect to (7.1), of an element of $\mathcal{V}_b(X)$ is bounded leads to the conclusion that such an exact b-metric is actually a smooth fibre metric on the vector bundle ${}^b TX$.[6]

That $\mathcal{V}_b(X)$ is a Lie algebra makes the 'enveloping algebra' $\mathrm{Diff}_b^*(X)$ a natural object filtered by order, i.e. an element $P \in \mathrm{Diff}_b^m(X)$ is an operator on $\mathcal{C}^\infty(X)$ which can be written as a finite[7] sum of products with up to m factors, each in $\mathcal{V}_b(X)$, or $\mathcal{C}^\infty(X)$. Thus, locally P is just a differential operator which can be written in the form

$$(7.5) \qquad Pu = \sum_{k+|\alpha|\leq m} p_{k,\alpha}(x,y)(xD_x)^k D_y^\alpha u$$

where the coefficients are $\mathcal{C}^\infty$ functions of the local coordinates, $xD_x = -ix\partial/\partial x$ and $D_y = -i(\partial/\partial y_1,\ldots,\partial/\partial y_p)$. Since this space of operators is clearly invariant under conjugation by a non-vanishing smooth function the corresponding space $\mathrm{Diff}_b^m(X;E)$ is defined for any vector bundle E. Let ${}^b\Lambda^k(X)$ be the k-fold exterior power of ${}^b T^*X$, the dual bundle to ${}^b TX$. Then

Lemma 7.1 [8] *The Laplacian of a(n exact) b-metric acting on forms on the interior of X extends by continuity to an element $\Delta \in \mathrm{Diff}_b^2(X; {}^b\Lambda^k)$ for every k.*

A more precise statement of the form of the Laplacian near the boundary can be obtained in terms of the orthogonal decomposition of the bundle

$$(7.6) \qquad {}^b T_{\partial X}^* X = (\frac{dx}{x})\mathbb{R} \oplus T^*\partial X$$

which follows from the form, (7.1), of the metric. Then the Laplacian,

---

[6] Not all (smooth) fibre metrics on ${}^b TX$ are exact b-metrics; this indeed is where the 'exactness' comes in. At a boundary point the element $x\partial/\partial x \in {}^b T_p X$ is completely natural. The first condition on an exact b-metric (as opposed to a general fibre metric on ${}^b TX$) is that $x\partial/\partial x$ should have length 1 at each boundary point. Then consider the subspace $W \subset \mathcal{V}_b(X)$ consisting of those element with $W_p \perp x\partial/\partial x$ at each boundary point. For an exact b-metric this is a Lie algebra (under commutation); in general it is not. A b-metric with this additional property could well be called a 'closed' b-metric. Restriction to the boundary gives a natural map $W \longrightarrow \mathcal{C}^\infty(\partial X; T\partial X)$ and the exactness of the metric, in the sense that it is of the form (7.1) in some product decomposition, is equivalent to the further global condition at ∂X that this should have a right inverse which is a Lie algebra homomorphism.

[7] Since X is compact, otherwise as a locally finite sum.

[8] This is easy enough to prove but notice that, for general k, it is not true if ${}^b\Lambda^k(X)$ is replaced by the usual form bundles $\Lambda^k(X)$.

on k-forms, becomes[9]

(7.7)
$$\Delta = \begin{pmatrix} (xD_x)^2 + \Delta_{\partial,k-1} & 0 \\ 0 & (xD_x)^2 + \Delta_{\partial,k} \end{pmatrix} + xQ, \ Q \in \mathrm{Diff}_b^2(X; {}^b\Lambda^k),$$

$${}^b\Lambda^k(X) = \frac{dx}{x} \wedge \Lambda^{k-1}(\partial X) \oplus \Lambda^k(\partial X) \ \text{near} \ \partial X.$$

Here $\Delta_{\partial,k}$ is the Laplacian on the boundary, for the metric h, acting on k-forms.

7.2 Thresholds

As already noted, the spectral and scattering theory for the Laplacian of an exact b-metric is dominated by the approximate splitting of the metric as a product. In particular the eigenvalues of the Laplacian on the boundary play an important role in the spectral and scattering theory of the Laplacian on X. For the Laplacian acting on k-forms the thresholds are the eigenvalues of the Laplacian on the boundary, acting on both k and $k-1$ forms.[10] The spectrum of Δ acting on k forms consists of branches of continuous spectrum starting at each threshold with finite multiplicity equal to the rank of the eigenspace(s) of the boundary Laplacian(s) corresponding to that threshold; in addition there may be L^2 eigenfunctions embedded in the continuous spectrum (and even in some cases below it). The L^2 eigenvalues necessarily form a discrete set and each is of finite multiplicity.

For a product-type metric Christiansen and Zworski have given optimal estimates on the growth rate of the counting function for the L^2 eigenvalues. Let $N_{\mathrm{pp}}(\lambda)$ be the number, counted with multiplicity, of L^2 eigenfunctions[11] with eigenvalues less than or equal to λ^2.

---

[9] In case of a product-type metric this decomposition becomes exact in a neighbourhood of the boundary, i.e. with $Q = 0$.

[10] For $k = 0$ or $k = \dim X$ this means just acting on 0 forms and $(\dim X - 1)$-forms respectively.

[11] Any eigenfunction of the Laplacian which is square integrable with respect to the metric inner product and volume form necessary decays at a uniform rate, depending on the eigenvalue, near the boundary,

(7.8) $\quad (\Delta - \lambda^2)u = 0, \ \int_X |u|^2 dg < \infty \Longrightarrow |u| \leq C_a x^a, \ a > 0,$

where a is determined by the nearest eigenvalue of the boundary Laplacian, i.e. threshold, to the eigenvalue. One can even see that such an eigenfunction has a complete asymptotic expansion in powers of x (with logarithmic factors).

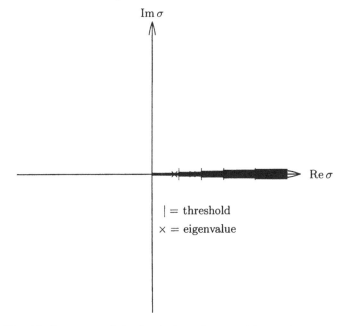

Fig. 12. Spectrum of the Laplacian of an exact b-metric.

Proposition 7.1 [12] *For a product-type metric*

$$(7.10) \qquad\qquad N_{\mathrm{pp}}(\lambda) \le C(1 + |\lambda|)^{\dim X}.$$

The behaviour of the continuous spectrum and the related definition of the scattering matrix can be obtained, as in the case of Euclidean scattering, by analytic continuation of the resolvent.

Theorem 7.1 [13] *The resolvent $(\Delta - \lambda^2)^{-1}$ extends, as an operator from $\dot{C}^\infty(X; {}^b\Lambda^k)$ to $C^{-\infty}(X; {}^b\Lambda^k)$, to a meromorphic function of $\lambda \in D$ where D is the minimal Riemman surface to which all the functions $(\lambda^2 - \tau_j{}^2)^{\frac{1}{2}}$ extend to be holomorphic, as $\tau_j{}^2$ runs over the set of thresholds.*

---

[12] The first result of this general type is due to Donnelly, [18], who showed that

$$(7.9) \qquad\qquad N_{\mathrm{pp}}(\lambda) \le C(1 + |\lambda|)^{2 \dim X - 1}.$$

The improvement of this estimate to (7.10) is due to Christiansen and Zworski, [16]; their proof uses a form of the 'pseudo-Laplacian' of Lax-Phillips and Colin de Verdière, which in turn is very closely related to the Atiyah-Patodi-Singer boundary condition discussed below.

[13] In fact one can be much more precise about the structure of the Schwartz kernel of the analytic continuation of the resolvent, see [73].

7.3 Scattering matrix

From the limiting absorption principle, i.e. an examination of the limit
of the resolvent as the parameter approaches the spectrum, a parametrization of the generalized eigenspaces can be obtained much as in the
Euclidean case.

Proposition 7.2 *If $\tau_j{}^2$ is an eigenvalue of the boundary Laplacian and
ϕ is an associated eigenfunction, $\Delta_\partial \phi = \tau_j{}^2 \phi$, then for each $\lambda \in \mathbb{R}$
with $\lambda^2 > \tau_j{}^2$ such that λ^2 is not a threshold [14] there exists a unique
eigenfunction, with eigenvalue λ^2, of the Laplacian satisfying[15]*

$$(7.11) \quad u = x^{i\sqrt{\lambda^2 - \tau_j{}^2}} \phi + \sum_{\tau_k{}^2 < \lambda^2} x^{-i\sqrt{\lambda^2 - \tau_j{}^2}} \psi_k + u', \; u' \in L_b^2(X)$$

$$(7.12) \qquad and \; u \perp v \; \forall \; v \in L_b^2(X) \; with \; (\Delta - \lambda^2)v = 0.$$

The sum in (7.11) is over the thresholds smaller than λ^2 and the
coefficients, ψ_k, are associated eigenfunctions. Let

$$(7.13) \qquad \text{Eig}(\lambda) = \sum_{\tau_k < \lambda^2} \text{null}(\Delta_\partial - \tau_k{}^2)$$

be the direct sum of the eigenfunctions for the boundary Laplacian with
eigenvalue less than λ^2. The coefficients ψ_k in (7.11) are determined by
ϕ so the map

$$(7.14) \qquad \phi \longmapsto \bigoplus_k (\frac{\sqrt{\lambda^2 - \tau_k{}^2}}{\sqrt{\lambda^2 - \tau_j{}^2}} \psi_k) \in \text{Eig}(\lambda)$$

extends by linearity to a map

$$(7.15) \qquad A(\lambda) : \text{Eig}(\lambda) \longrightarrow \text{Eig}(\lambda).$$

This is the scattering matrix, [16] at frequency λ, for the metric. The
λ-dependent factors are included to make the scattering matrix unitary.
Note that it is a matrix, at least in the sense of an operator on a finite
dimensional space, with the caveat that the dimension of the space in-
creases every time λ^2 crosses a threshold, the change being the dimension
of the associated eigenspace of the boundary Laplacian.[17]

---

[14] i.e. is not an eigenvalue of the boundary Laplacian.

[15] Here $L_b^2(X)$ is the L^2 space for any b-metric on X; the volume form is $\frac{dx}{x}dy$
near the boundary. Since the generalized eigenfunction in (7.11) is bounded and
L^2 eigenfunctions decay, as stated in Footnote 11, the pairing condition in (7.12)
does make sense.

[16] As normalized by Christiansen in [15]; the normalization in [73] is different.

[17] The results here are stated for the Laplacain on functions. For the Laplacian
on forms essentially the same result holds, except that the 'boundary Laplacian'

7.4 Boundary expansions and pairing

To see the unitarity of $A(\lambda)$, and other things besides, it is convenient to consider a limiting form of Green's theorem in this setting. For any $\lambda \in \mathbb{R}$ suppose $u \in \mathcal{C}^{-\infty}(X; {}^b\!\Lambda^k)$[18] and that u satisfies $(\Delta - \lambda^2)u \in \dot{\mathcal{C}}^\infty(X; {}^b\!\Lambda^k)$. Then, for product-type metrics, and for λ^2 not a threshold, u necessarily has a complete asymptotic expansion at the boundary in the sense that

$$(7.17) \qquad u \sim \sum_j x^{\mu^\pm(j)} \phi_{j,\pm}, \ \ \mu^\pm(j)^2 = \tau_j^2 - \lambda^2,$$

where $\phi_{j,\pm}$ are eigenfunctions, with eigenvalue τ_j, for the boundary Laplacian. Notice that the assumption that the solutions are of polynomial growth means that there can only be a finite number of terms with negative powers of x.[19] If $\lambda^2 = \tau_j^2$ for some threshold then there is just one corresponding $\mu(j) = 0$ and in place of (7.17) one must allow the possibility of a logarithmic term:

$$u = \phi_{j,0} + \log x \cdot \phi_{j,1} + \sum_{p \neq j} x^{\mu^\pm(p)} \phi_{p,\pm},$$

$$(7.18)$$

$$\mu^\pm(p)^2 = \tau_p^2 - \tau_j^2, \ \lambda^2 = \tau_j^2.$$

Here the $\phi_{p,\pm}$ are eigenfunctions as before[20] and $\phi_{j,i}$ are eigenfunctions for the boundary Laplacian with eigenvalue τ_j^2.[21]

should be interpreted as the matrix of operators arising from (7.7), i.e.

$$(7.16) \qquad \begin{pmatrix} \Delta_{\partial,k-1} & 0 \\ 0 & \Delta_{\partial,k} \end{pmatrix}.$$

---

[18] This just means that u can be written as (a finite sum of sections of the form bundle of the type) $x^{-N} B_N v$ where v is square-integrable and $B_N \in \operatorname{Diff}_b^N(X; {}^b\!\Lambda^k)$ for some N. Thus u is of at worst 'polynomial growth' near the boundary.

[19] Hence there can only be a finite number of terms with power having real part less than any given real number p. It follows that the asymptotic expansion makes sense.

[20] See Footnote 17.

[21] For a general exact b-metric something very similar is true. There is an asymptotic expansion (7.17), or (7.18), as λ^2 is not, or is, a threshold, but the coefficients $\phi_{j,\pm}$ cannot be taken to be constant. In fact they may not even be smooth but are necessarily of the form

$$(7.19) \qquad \phi_{j,\pm} \sim \sum_{r=0}^{\infty} x^{N(r,j)} (\log x)^r \phi_{j,\pm,r}.$$

Here the coefficients $\phi_{j,\pm,r}$ are all smooth and $N(r,j) \to \infty$ as $r \to \infty$ for each j, so the sum makes sense asymptotically. The leading parts $\phi_{j,\pm,0} \restriction (x = 0)$ are eigenfunctions as before. The logarithmic terms in (7.19) arise only because of

The boundary pairing is mostly of interest for 'extended L^2 solutions.' Thus, consider the space which is just a little larger[22] than L^2

(7.20) $$x^{0-}L_b^2(X;{}^b\Lambda^k) = \bigcap_{\epsilon>0} x^{-\epsilon}L_b^2(X;{}^b\Lambda^k).$$

Proposition 7.3 *If λ^2 is not a threshold and $u^{(i)} \in x^{0-}L_b^2(X;{}^b\Lambda^k)$ for $i = 1,2$ satisfy $(\varDelta - \lambda^2)u^{(i)} \in \dot{\mathcal{C}}^{\infty}(X;{}^b\Lambda^k)$ then*[23]

(7.21)
$$\int_X \left(\langle(\varDelta - \lambda^2)u^{(1)}, u^{(2)}\rangle - \langle u^{(1)}, (\varDelta - \lambda^2)u^{(2)}\rangle \right) dg$$
$$= \sum_{\tau_j^2 < \lambda^2} 2i\sqrt{\lambda^2 - \tau_j^2} \int_{\partial X} \left(\langle \phi_{j,+}^{(1)}, \phi_{j,+}^{(2)}\rangle - \langle \phi_{j,-}^{(1)}, \phi_{j,-}^{(2)}\rangle \right) dh.$$

If $\lambda^2 = \tau_k^2$ is a threshold then the right side of (7.21) is replaced by

(7.22)
$$2i \int_{\partial X} \left(\langle \phi_{k,1}^{(1)}, \phi_{k,0}^{(2)}\rangle - \langle \phi_{k,0}^{(1)}, \phi_{k,1}^{(2)}\rangle \right) dh$$
$$+ \sum_{\tau_j^2 < \tau_k^2} 2i\sqrt{\tau_k^2 - \tau_j^2} \int_{\partial X} \left(\langle \phi_{j,+}^{(1)}, \phi_{j,+}^{(2)}\rangle - \langle \phi_{j,-}^{(1)}, \phi_{j,-}^{(2)}\rangle \right) dh.$$

7.5 Hodge theory

It turns out that the Hodge cohomology [24] of an exact b-metric is the same as that of a scattering metric:

Proposition 7.4 *The null space of the Laplacian, acting on square-integrable k-forms, is naturally isomorphic to the image of the relative (deRham) cohomology of X in the absolute cohomology.*

In fact quite a lot more can be said about 'nearly square-integrable' solutions.[25] Consider the 'extended L^2 null space.' As a special case

accidental multiplicities, when two of the $\mu^{\pm}(j)$ differ by an integer. In case λ^2 is a threshold a similar modification is needed to (7.18).

[22] Here L_b^2 is just the metric L^2 space, in fact independent of choice of b-metric.

[23] Here I have only written the formulæ in the case of a product-type metric. In the general case the result is the same, except for instance that the terms on the right in (7.21) are the leading coefficients in (7.19).

[24] The modifications required in the discussion above when the boundary of X has more than one component are rather minor. The boundary operator should just be interpreted as the direct sum of the operators on the different components and sums, as in (7.21), are over the eigenvalues of all these operators.

[25] See Chapter 5 of [73] for a more detailed discussion.

of (7.18) or (7.19), if $u \in x^{0-} L_b^2(X; {}^b\Lambda^k)$ is in the null space of the Laplacian then near the boundary it takes the form

(7.23)
$$u = (u_{10} + u_{11} \log x) \wedge \frac{dx}{x} + (u_{00} + u_{01} \log x) + u', \ u' \in L_b^2(X; {}^b\Lambda^k)$$

where the u_{ij} are $k - i$ forms on ∂X, independent of x, which are in the null space of the boundary Laplacian, i.e. they are harmonic.

The range of this generalized boundary data, given by the u_{ij}'s, is a Lagrangian subspace of the symplectic vector space formed by the direct sum of two copies of the Hodge cohomology of ∂X in dimensions $k - 1$ and k. Moreover, both the relative and the absolute cohomology of X can be 'found' in this extended L^2 null space:

Theorem 7.2 *The relative cohomology of X is naturally identified with the space of harmonic forms in $x^{0-} L_b^2(X; {}^b\Lambda^k)$ such that $u_{00} = 0$, $u_{01} = 0$ and $u_{11} = 0$ in (7.23), and similarly the absolute cohomology is naturally identified with the space of harmonic forms in $x^{0-} L_b^2(X; {}^b\Lambda^k)$ such that $u_{00} = 0$, $u_{10} = 0$ and $u_{11} = 0$. All these forms are closed and coclosed.*

7.6 Atiyah-Patodi-Singer index theorem

A b-differential operator $P \in \text{Diff}_b^m(X; E, F)$, acting from sections of one vector bundle E to those of another bundle, F, has an invariantly defined principal symbol,

(7.24) $\sigma_{b,m}(P) \in \mathcal{C}^\infty({}^bT^*X; \pi^* \text{hom}(E, F))$.

Here $\pi : {}^bT^*X \longrightarrow X$ is the projection and $\sigma_{b,m}(P)$ is a homogeneous polynomial of degree m on the fibres of ${}^bT^*X$. As in the last lecture I prefer to replace ${}^bT^*X$ by its stereographic compactification ${}^b\bar{T}^*X$.[26] The new part added to compactify the vector bundle is just ${}^bS^*X$, the sphere bundle of ${}^bT^*X$. Then the symbol map (7.24) can be regarded as a map

(7.25) $\sigma_{b,m}(P) \in \mathcal{C}^\infty({}^bS^*X; \pi^* \text{hom}(E, F) \otimes Q_m)$.

where Q_m is the bundle over ${}^bS^*X$ whose sections are the functions which are homogeneous of degree m on ${}^bT^*X$. The differential operator

---

[26] This is a compact manifold with corner; it is actually diffeomorphic to ${}^{sc}\bar{T}^*X$, just not naturally so.

is elliptic,[27] as an element of $\mathrm{Diff}_b^m(X; E, F)$ if $\sigma_{b,m}(P)(\zeta)$ is, for each $\zeta \in {}^bT^*X \setminus 0$, invertible as a map from $E_{\pi(\zeta)}$ to $F_{\pi(\zeta)}$. The symbol of P in (7.5) is just

$$(7.26) \qquad \sigma_{b,m}(P) = \sum_{k+|\alpha|=m} p_{k,\alpha}(x,y)\lambda^k\eta^\alpha.$$

As well as this symbol there is a closely related 'non-commutative boundary symbol.' In fact this 'indicial operator,'

$$(7.27) \qquad I(P,\lambda) = \sum_{k+|\alpha|\leq m} p_{k,\alpha}(0,y)\lambda^k D_y^\alpha,$$

is an element of $\mathrm{Diff}^m(\partial X; E_{\partial X}, F_{\partial X})$ which depends polynomially on $\lambda \in \mathbb{C}$. It is invariantly defined, except that it depends mildly on the choice of boundary defining function, x.

The b-Sobolev spaces, of positive integral order, are defined quite analogously to the usual ones:

$$(7.28) \quad H_b^k(X; E) = \left\{ u \in L_b^2(X; E); \mathrm{Diff}_b^k(X; E)u \subset L_b^2(X; E) \right\}.$$

Theorem 7.3 [28] *An element $P \in \mathrm{Diff}_b^m(X; E, F)$, where X is a compact manifold with boundary, is Fredholm as an operator $P : H_b^m(X; E) \longrightarrow L_b^2(X; F)$ if and only if it is elliptic and has indicial operator $I(P, \lambda) : C^\infty(\partial X; E_{\partial X}) \longrightarrow C^\infty(\partial X; F_{\partial X})$ invertible for each real λ.*

If X is a spin manifold[29] the spin bundle, S, over X is defined and the Dirac operator associated to an exact b-metric is an elliptic element of $\mathrm{Diff}_b^1(X; S)$. If X is even-dimensional then $S = S^+ \oplus S^-$ splits as a direct sum of two subbundles and $\eth$ is '$\mathbb{Z}_2$-graded odd' i.e. decomposes as an anti-diagonal 2×2 matrix

$$(7.29) \qquad \eth = \begin{pmatrix} 0 & \eth^- \\ \eth^+ & 0 \end{pmatrix}.$$

Over the boundary, S^+ and S^- can be identified in a natural way and in terms of this identification the indicial family of $\eth^+$ is just the family $\eth_0 + i\lambda$, where $\eth_0$ is the (self-adjoint) Dirac operator on the boundary

---

[27] One should perhaps say 'b-elliptic' to distinguish this condition from ellipticity in the ordinary sense. As it is impossible for P to be elliptic in this stronger sense if $m > 0$ (and the two ellipticities are the same for $m = 0$) there should be no confusion between the two notions.

[28] This is a generalization of traditional results for ordinary differential operators with regular-singular points. It was proved in this generality in [75].

[29] For a discussion of these matters, and the Atiyah-Patodi-Singer theorem itself, see [73].

defined by the induced spin structure. The indicial family is therefore invertible for all real λ if and only if it is invertible for $\lambda = 0$, i.e. when $\eth_0$ itself is invertible. Thus Theorem 7.3 shows that $\eth^+ : H_b^1(X; S^+) \longrightarrow L_b^2(X; S^-)$ is Fredholm if and only if $\eth_0$ is invertible.

Theorem 7.4 [30] *If X is an even-dimensional compact spin manifold with boundary then $\eth^+$, the positive Dirac operator, is Fredholm if and only if the boundary Dirac operator $\eth_0$ is invertible; its index is then*[31]

$$\operatorname{ind}(\eth^+) = \dim \operatorname{null}(\eth^+) - \dim \operatorname{null}(\eth^-) = \int_X \widehat{A} - \frac{1}{2}\eta(\eth_0)$$

(7.30)

$$where \ \eta(\eth_0) = \frac{1}{\sqrt{\pi}} \int\limits_0^\infty t^{-\frac{1}{2}} \operatorname{Tr}\left(\eth_0 \exp(-t\eth_0^2)\right) dt.$$

7.7 b-Pseudodifferential operators

As in Section 6.5 the possibility of extending the Atiyah-Patodi-Singer index theorem to a general index theorem for b-differential, or even b-pseudodifferential, operators is of considerable interest. As far as I know this has not been done in a completely satisfactory way.[32] The algebra of pseudodifferential operators, $\Psi_b^*(X; E)$, which 'microlocalizes' the algebra $\operatorname{Diff}_b^*(X; E)$ is relatively straightforward to define. I shall simply remark on some of its properties; see [68], [44] (Section 18.3) or [73] for the definition.[33] Other closely related classes of operators have been defined by Kondratev[50], Plamenevskii [93] and Schulze [102].[34]

For any $m \in \mathbb{R}$ and any vector bundle E over X, $\Psi_b^\infty(X; E)$ consists of a linear space of operators on $\mathcal{C}^\infty(X; E)$ which is filtered by the order

---

[30] This is the Atiyah-Patodi-Singer index theorem for the Dirac operator for a spin structure. It has rather direct extensions to more general Dirac operators associated to Hermitian Clifford modules, as discussed, for example, in [73]. For more general 'family index theorems' of Atiyah-Patodi-Singer type see the work of Bismut and Cheeger [10], [77], [78].

[31] The $\widehat{A}$-genus here is the characteristic class, expressed in terms of the Riemann curvature by $\widehat{A} = \det^{\frac{1}{2}}\left(\frac{R/4\pi i}{\sinh(R/4\pi i)}\right)$. The convergence of the integral defining the η invariant of the boundary Dirac operator also requires justification.

[32] However, see the work of Piazza [92] for a partial result.

[33] See [76] for a brief discussion in the case of manifolds with corners.

[34] There is a difference between the calculus $\Psi_b^*(X; E)$ and the others listed here arising from the different intended applications, to complete manifolds on the one hand and incomplete manifolds with conical singularities on the other. For a general discussion of pseudodifferential operators on complete manifolds with bounded geometry see the lectures of Shubin [107].

m and

$$(7.31) \qquad \operatorname{Diff}_b^m(X; E) \subset \Psi_b^m(X; E) \ \forall \ m \in \mathbb{N}.$$

The symbol map is defined, consistently with (7.26) and this inclusion, as a map onto the space $\mathcal{C}^\infty({}^bS^*X; \pi^* \hom(E) \otimes Q_m)$ of sections which are homogeneous of degree m. This symbol map is multiplicative and gives a short exact sequence

$$(7.32) \qquad \sigma_{b,m+m'}(PQ) = \sigma_{b,m}(P) \cdot \sigma_{b,m'}(Q),$$

(7.33)
$$0 \longrightarrow \Psi_b^{m-1}(X; E) \longrightarrow \Psi_b^m(X; E) \xrightarrow{\sigma_{b,m}} \mathcal{C}^\infty({}^bS^*; \pi^* \hom(E) \otimes Q_m) \longrightarrow 0.$$

Furthermore the map to the indicial family, (7.27), also extends to the algebra of b-pseudodifferential operators; it takes values in the space of entire functions with values in $\Psi^m(\partial X; E_{\partial X})$ with appropriate uniformity properties in the complex parameter λ. Furthermore Theorem 7.3 extends to the pseudodifferential operators essentially *verbatim* to characterize the Fredholm elements of the algebra.

7.8 Trace formula and spectral asymptotics

As already noted above the Poisson trace formula has been extended, by Christiansen, to exact b-metrics. If $x \in \mathcal{C}^\infty(X)$ is a boundary defining function with respect to which the metric takes the form (7.1) and $U_g(t)$ is the wave group as in (4.2), but on X, then the b-trace can be defined, as a tempered distribution, by Hadamard regularization:

(7.34)
$$\tau_g(\rho) = \text{b-Tr}\, U(\rho) = \lim_{\epsilon \downarrow 0} \left(\int_{x > \epsilon} \operatorname{tr} \int_{\mathbb{R}} U(t, p, p) \rho(t) dt\, dg(p) - A(\rho) \log \epsilon \right),$$

where tr is just the trace on 2×2 matrices and $A(\rho)$ is determined by the condition that the limit exist.

Proposition 7.5 [35] *For any exact b-metric the inverse Fourier transform of the b-trace (7.34) of the wave group satisfies*

(7.35)

$$\mathcal{F}^{-1}\tau_g = \sum_{(\lambda')^2 \in \text{ppspec}(\Delta)} \delta(\lambda - \lambda') + \frac{1}{2\pi i}\frac{d}{d\lambda}\log\det A(\lambda)$$

$$+ \frac{1}{4}\sum_{0\neq\sigma^2\in\text{spec}(\Delta_0)}\delta(\lambda-\sigma) + \frac{1}{2}\operatorname{Tr}A(0)\delta(\lambda).$$

Using this trace formula Christiansen and Zworski have shown that the sum of the counting function for the point spectrum and the scattering phase has a leading 'Weyl' term as $\lambda \to \infty$:

(7.36)

$$N_{\text{pp}}(\lambda) + \frac{1}{2\pi i}\log\det A(\lambda) = (2\pi)^{-n}\operatorname{Vol}(\mathbb{B}^n)\text{b-}\operatorname{Vol}(X)\lambda^n + O(\lambda^{n-1})$$

$$\text{where b-}\operatorname{Vol}(X) = \int_X^b dg = \lim_{\epsilon\downarrow 0}\left(\int_{x>\epsilon}dg + \log\epsilon\operatorname{Vol}(\partial X)\right)$$

is the regularized 'b-volume' of the manifold.

7.9 Manifolds with corners

There are many interesting possible extensions of these results. In particular it is natural to seek similar descriptions of the index, spectral and scattering theory for similar exact b-metrics on manifolds with corners . These are easily defined inductively. Indeed a compact manifold with corners, X, has a product decomposition in a neighbourhood of each of its boundary hypersurfaces of the form $[0,\epsilon) \times H$ where the boundary hypersurface, H, is a compact manifold with corners where the maximum codimension of a boundary face is at most one less than in X. Thus if, for each boundary hypersurface H, x_H is a defining function for H then an exact b-metric of product type on X is a metric on the interior which takes the form

(7.37) $$g = (\frac{dx_H}{x_H})^2 + g'_H \text{ near } H$$

---

[35] The scattering phase, $\log\det A(\lambda)$ has to be normalized at each threshold, since the rank of $A(\lambda)$ changes there. The normalization of the jump of the phase used here, following [15], involves the behaviour of the generalized eigenfunctions. This is important since it contributes δ terms to the formula.

where g'_H is an exact b-metric of product type on H. It is rather clear that one should expect the following.[36]

Conjecture 7.1 *For an exact b-metric[37] the spectrum of the Laplacian consists of possibly countable point spectrum of finite multiplicity, corresponding to L^2 eigenfunctions together with continuous spectrum consisting of an at most countable union of rays $[\tau, \infty)$ where the thresholds, τ, form the discrete set consisting of the union of the L^2 eigenvalues for the induced Laplacians on all boundary faces (of positive codimension).[38]*

Notice that Euclidean spaces are themselves examples of these types of manifolds. So are products of manifolds with corners, carrying exact b-metrics. In both these cases this conjecture is easily verified.

It would also be interesting to extend the Atiyah-Patodi-Singer index theorem to Dirac operators on such manifolds. At this stage there seems to be no general result except under the (unreasonable) assumption that the induced Dirac operator on every boundary face of codimension greater than 1 is invertible. Then a result very similar to Theorem 7.4 holds.[39]

---

[36] I hope to prove this soon!

[37] Whether of product type or not.

[38] In fact one should expect to be able to show that the ray of continuous spectrum arising from an L^2 eigenvalue on a boundary face of codimension p corresponds to functions on a (pseudo-) manifold of dimension $p - 1$. Part of the problem is to know exactly what is meant by a pseudomanifold here!

[39] Partial results on the behaviour of the eta invariant for the boundary of a manifold with corners up to codimension two are contained in [60] and [35].

8
Hyperbolic metrics

In the previous two lectures I discussed two classes of complete metrics on the interiors of compact manifolds with boundaries. Today I will discuss two more,[1] necessarily in less detail. To start with let me briefly consider the classification of metrics which are 'warped products' near the boundary.

8.1 Warped products

Thus, if X is a manifold with boundary and x is a boundary defining function consider a metric which takes the form

$$(8.1) \qquad g = x^{2a}dx^2 + x^{2b}h$$

where at least initially it can be supposed that h is a metric on the boundary pulled back under some product decomposition, i.e. is independent of x and has no dx terms. Such a metric is complete[2] if and only if

$$(8.3) \qquad a \le -1.$$

---

[1] In a sense three more.

[2] A metric is complete if and only if every geodesic can be continued indefinitely in both directions. If X is compact then the only failure of completeness that can occur must involve a geodesic which approaches the boundary. Certainly condition (8.3) is necessary since the curves which are constant in the boundary factor are geodesics and these have infinite arclength in the direction in which x is decreasing if and only if

$$(8.2) \qquad \int_0^1 x^{-a}dx = \infty$$

which is just (8.3). The converse involves a closer look at the geodesic equations but is not very difficult.

I shall call the case $a = -1$ 'marginally complete.'[3]

In the marginally complete case the sign of b is the determining feature. Indeed, for any constant $q > 0$ the transformation $x' = x^q$, whilst not smooth, is continuous and the introduction of x^q as a boundary defining function simply changes the C^∞ structure on X near the boundary. Thus in these cases, and even if h involves both dx and x dependence of the coefficients, the distinguished cases[4] are[5]

$$(8.4) \qquad g = \frac{dx^2}{x^2} + x^2 h, \quad \text{'hc'=hyperbolic, or rank one, cusp}$$

$$(8.5) \quad g = \frac{dx^2}{x^2} + h, \quad \text{'b'=boundary or metric with cylindrical end}$$

$$(8.6) \quad g = \frac{dx^2}{x^2} + \frac{h}{x^2}, \quad \text{'0'=zero or conformally compact metric.}$$

The first and last of these are the cases I wish to discuss today. The middle one is of course the subject of Lecture 7.

Now consider the 'overcomplete' cases, where $a < -1$. There is a basic classification corresponding to the sign of $b - a - 1$. If this vanishes then the, possibly singular, transformation $x \longmapsto x' = x^{-1/(1+a)}$ reduces the metric to a constant multiple of the scattering metric

$$(8.7) \qquad g_{sc} = x^{-2} \left(\frac{dx^2}{x^2} + h \right) = \frac{dx^2}{x^4} + \frac{h}{x^2}, \quad \text{'sc'=scattering,}$$

considered in Lecture 6. If $b < 1 + a$ (which is, by assumption, negative) then the transformation $x' = x^{-b/(1+a)}$ reduces the metric to a constant multiple of the singular conformal 0-metric

$$(8.8) \qquad g = x^{-2r} \left(\frac{dx^2}{x^2} + \frac{h}{x^2} \right), \quad r = \frac{(1+a)^2}{b} > 0.$$

In the remaining cases, where $b > 1 + a$, the metric is conformal, with a possibly singular conformal factor, to the 'hyperbolic cusp' type metric in (8.4). It is worth noting the three subcases. First if $(1 + a) < b < 0$

---

[3] One should be a little careful about 'degrees of completeness' since this is rather subjective. See Footnote 6.

[4] There does not seem to be any uniformity in the names or notation for these various types of metrics. I like to have abbreviations, such as '0-' in order to be able to denote in a simple way the various objects that are 'functorially' associated to these types of metrics, e.g. the '0-pseudodifferential operators' discussed below. A rather more complete discussion of *how* one can associate such objects with these classes of metrics (and many others, a couple more of which are mentioned below) is to be contained in [66]. Maybe it exists by now?

[5] Here and below I consistently switch to the old notation, x, for the new boundary defining function.

then the metric can be transformed to a constant multiple of

$$(8.9) \qquad g = x^{-2r}\left(\frac{dx^2}{x^2} + x^2 h\right) \text{ with } r > 1.$$

In case $b = 0$ again a, possibly singular, transformation $x \longmapsto x' = x^{-1-a}$ can be made reducing the metric to a constant multiple of the case[6]

$$(8.10) \qquad g_c = \frac{dx^2}{x^4} + h \quad \text{'c'=cusp}.$$

In the remaining case of $b > 0$ a similar transformation leads to the remaining case intermediate between (8.4) and (8.10)

$$(8.11) \qquad g = x^{-2r}\left(\frac{dx^2}{x^2} + x^2 h\right) \text{ with } 0 < r < 1.$$

In summary then there are eight basic cases. I do not know of any serious examination of (8.8), (8.9) or (8.11). The other cases have all been treated to a greater or lesser extent. A similar classification can be made in the incomplete cases.

8.2 Conformally compact manifolds

The first case I shall discuss is a metric of the form (8.6) near the boundary. For the most part the assumption that h is independent of x and dx can be dropped since it only affects the more subtle results.

The basic example of such a metric is that on hyperbolic space. In the upper-half space model, n-dimensional[7] hyperbolic space is just[8]

$$(8.12) \qquad \mathbb{H}^n = \left\{(x,y) \in \mathbb{R}^n; x > 0, \ y \in \mathbb{R}^{n-1}\right\}$$

with the metric

$$(8.13) \qquad g = \frac{|dx|^2 + |dy|^2}{x^2}.$$

Clearly this is of the form (8.6) down to $x = 0$. Of course this is not a

---

[6] Notice that the transcendental transformation $x \longmapsto x' = 1/\log(1/x)$ reduces the cusp metric in (8.10) to the metric (8.5). Of course if one has a variable coefficient form of (8.10), where h depends on x then this introduces rather singular coefficients in (8.5), so one should really think of smooth perturbations of (8.5) as a special case of smooth perturbations of (8.10).

[7] $n \geq 2$.

[8] In most discussions of hyperbolic space the role of the variables in (8.12) is reversed, i.e. the boundary defining functions is denoted y and the x's are variables in $\mathbb{R}^{n-1}$. I think it is better to keep to the consistent usage of x as a boundary defining function.

compact manifold with boundary but it can be compactified to a Euclidean ball so that the hyperbolic metric is a 0-metric. I shall denote the compact model by $\bar{\mathbb{H}}^n$.

There are other examples of 0-metrics which arise from hyperbolic space. The group of diffeomorphism of $\mathbb{H}^n$ preserving the metric is $\mathrm{PSL}(n, \mathbb{R})$, acting as fractional linear transformations. Consider Γ which is a subgroup acting discretely.[9] If Γ has no fixed points (in the interior) then $\mathbb{H}^n/\Gamma$ is a manifold, although it can be rather nasty. If the group Γ is 'small' in the sense that it has no parabolic elements and is geometrically finite[10] then the quotient[11] $\bar{\mathbb{H}}^n/\Gamma$ is again a compact manifold with boundary to which the hyperbolic metric descends as a 0-metric. The extension of scattering theory to hyperbolic space, and its quotients, has quite a long history; in particular Lax and Phillips extended their theory to this setting; see especially [54] and references therein.

The metric on hyperbolic space is very special in that its sectional curvature is a negative constant. A general 0-metric is not so restricted. Let me write (8.6) in the form

$$(8.14) \qquad g = \frac{g_\infty}{x^2}$$

where g_∞ is an incomplete metric which is smooth up to the boundary, i.e. is just a smooth positive definite metric on the fibres of TX.[12]Even so such a 0-metric has 'asymptotically constant curvature' in the weak sense that if $p \in \partial X$ and $p_n \in X^\circ$ is a sequence with limit p then the sectional curvatures at p_n approach the fixed value $-|dx|_\infty^2$, where $|\cdot|_\infty$ is the dual metric on T^*X to the metric g_∞. This suggests a strengthening of (8.14) by adding the condition

$$(8.15) \qquad |dx|_\infty = 1 \text{ on } \partial X.$$

This implies that, for an appropriate choice of product structure,

$$(8.16) \qquad g = \frac{dx^2 + h(x)}{x^2} \text{ near } \partial X,$$

where h is a metric on the cross-section of the product, i.e. on ∂X but depending parametrically on x. This can be further restricted by

---

[9] Meaning that for any compact subset, K, of the interior of $\mathbb{H}^n$ the set of $\gamma \in \Gamma$ such that $\gamma(K) \cap K \neq \emptyset$ is finite.

[10] For a full discussion of this condition see the paper of Bowditch [11].

[11] Examples are given by the Schottky groups.

[12] Alternatively this is the same as demanding that g_∞ be the restriction to X of a Riemann metric on $2X$, its double to a compact manifold without boundary.

demanding that $h(x)$ be independent of x :

(8.17) $$g = \frac{dx^2 + h}{x^2} \text{ near } \partial X.$$

Any compact manifold with boundary has a metric of this form.

8.3 0-geometry and analysis

By 0-geometry I mean the geometry of a conformally compact metric, (8.14), or more generally the geometry of the more fundamental object, $\mathcal{V}_0(X)$, which is the space of smooth vector fields on X of bounded length with respect such a metric. These structures were first discussed in this degree of generality by Mazzeo [59] and then in [61]. The space $\mathcal{V}_0(X)$ is independent of the particular 0-metric chosen

(8.18) $\mathcal{V}_0(X) = \{V \in \mathcal{V}(X); |V|_\infty \leq Cx\} = x\mathcal{V}(X) \subset \mathcal{V}_b(X).$

It consists of the vector fields which vanish[13] at the boundary. Again it is a Lie algebra. It has certain other properties[14] which mean that one can follow through much of the abstract discussion of differential operators on compact manifolds without boundary. Thus, as in the cases of the sc and b-structures, the filtration of the enveloping algebra of $\mathcal{V}_0(X)$, $\mathrm{Diff}_0^m(X)$, is defined, as consisting of those operators

(8.19) $$P : \mathcal{C}^\infty(X) \longrightarrow \mathcal{C}^\infty(X)$$

which can be written as a sum of up to m fold products of vector fields in $\mathcal{V}_0(X)$.[15] This definition can be extended, by localization, to define the filtered algebra $\mathrm{Diff}_0^*(X; E)$ of operators on sections of any vector bundle over X.

There is also a natural 0-tangent bundle, ^0TX. This can be defined either by 'rescaling,' so it can be denoted xTX or, as in Footnote 7.4, more abstractly so that

(8.20) $$\mathcal{V}_0(X) = \mathcal{C}^\infty(X; {}^0TX).$$

Its dual bundle, $^0T^*X$, plays the rôle of T^*X for these operators. In particular the standard symbol map extends by continuity from the interior to a map

(8.21) $\sigma_{0,m} : \mathrm{Diff}_0^m(X; E) \longrightarrow \mathcal{C}^\infty({}^0S^*X; \pi^* \hom(E) \otimes Q_m).$

---

[13] Hence the '0' moniker.
[14] I collect such properties together in the notion of a 'boundary fibration structure' discussed in [66].
[15] The fact that $\mathcal{V}_0$ is a Lie algebra makes the order well defined.

Here $^0S^*X$ is the sphere bundle over X with fibre at each point the sphere at infinity[16] and Q_m is the restriction to $^0S^*X$ of the line bundle with sections the functions on $^0T^*X$ which are homogeneous of degree m on the fibres. The null space of (8.21) is precisely $\text{Diff}_0^{m-1}(X;E)$.

As in the other cases of such rings of differential operators discussed above, this symbol map does not fix the important properties[17] of such operators. The second 'symbol' map is, as in the b-case, non-commutative. In fact it is 'partly commutative' and partly non-commutative. To describe it precisely observe that, for each $p \in \partial X$, ^0T_pX has a non-trivial Lie algebra structure. Namely if $v, w \in {}^0T_pX$ and $V, W \in \mathcal{V}_0(X)$ are such that $V(p) = v$ and $W(p) = w$[18] then

$$(8.22) \qquad [v,w] = [V,W](p) \in {}^0T_pX$$

is independent of the choice of V and W.[19] This makes ^0T_pX into a solvable Lie algebra[20] which is given in terms of the basis

$$x\frac{\partial}{\partial x}, \; x\frac{\partial}{\partial y_j}, \; j = 1, \ldots, n-1, \; \text{by}$$

(8.24)
$$[x\frac{\partial}{\partial x}, x\frac{\partial}{\partial y_j}] = -[x\frac{\partial}{\partial y_j}, x\frac{\partial}{\partial x}] = x\frac{\partial}{\partial y_j}, \; j = 1, \ldots, n-1$$

with all other commutators zero. Let $\text{Diff}_{0,p}^m(^0T_pX)$ be the part of order m of the enveloping algebra of this Lie algebra.[21] It is a finite-dimensional vector space and these spaces form a bundle of algebras over the boundary; I shall denote by $\text{Diff}_{I,0}^m(\partial X)$ the smooth sections of this bundle. The normal map is a surjective linear map

$$(8.25) \qquad N : \text{Diff}_0^m(X) \longrightarrow \text{Diff}_{I,0}^m(\partial X)$$

---

[16] This is just the boundary of the stereographic compactification of $^0T_p^*X$.

[17] For instance the Fredholm condition.

[18] In the sense of sections of ^0TX.

[19] This independence of choice is just the fact that if $\mathcal{I}(p) \subset \mathcal{C}^\infty(X)$ is the ideal of functions vanishing at $p \in \partial X$ then $\mathcal{V}_0(X) \cdot \mathcal{I}(p) \subset \mathcal{I}(p)$, so

$$(8.23) \qquad [\mathcal{V}_0(X), \mathcal{I}(p) \cdot \mathcal{V}_0(X)] \subset \mathcal{I}(p) \cdot \mathcal{V}_0(X)$$

shows that $\mathcal{I}(p) \cdot \mathcal{V}_0(X) \subset \mathcal{V}_0(X)$ is a Lie algebra ideal. The quotient $\mathcal{V}_0(X)/(\mathcal{I}(p) \cdot \mathcal{V}_0(X))$, which is, by definition, just ^0T_pX, is therefore a Lie algebra.

[20] So the corresponding Lie group is diffeomorphic to the same space, ^0T_pX, as the Lie algebra and the group law can be written down quite explicitly.

[21] If the Lie algebra ^0T_pX is realized as the left-invariant vector fields on the corresponding Lie group then $\text{Diff}_{0,p}^m(^0T_pX)$ consists of the operators on $\mathcal{C}^\infty$ functions on the Lie group which are polynomials, with constant coefficients and degree at most m, in these vector fields.

which is a homomorphism of algebras.[22]

I shall not go into any detail here, but part of the package of a 'boundary fibration structure' is that there is an associated ring of pseudodifferential operators, in this case the 0-pseudodifferential operators $\Psi_0^{*,*}(X; E)$, acting on sections of any given vector bundle, E. It has properties closely related to those of $\text{Diff}_0^*(X; E)$. The first of the orders in $\Psi_0^{*,*}(X; E)$ is the interior order, as a pseudodifferential operator in the ordinary sense, and the second order represents the degree of vanishing, or singularity, at the boundary. If x is a boundary defining function then

$$(8.27) \qquad \Psi_0^{m,l}(X; E) = x^l \Psi_0^{m,0}(X; E) \; \forall \; m, l \in \mathbb{R}.$$

Both the symbol map and the normal operator have natural extensions to these spaces of pseudodifferential operators.

The Lie group with Lie algebra $^0 T_p X$, $p \in \partial X$, has a natural compactification, $\bar{G}_p$, to a half-ball on which the elements of the Lie algebra are smooth vector fields. Acting on the natural 0-Sobolev spaces, as they do, an element $P \in \text{Diff}_0^m(X; E)$, or $\Psi_0^{m,l}(X; E)$, is Fredholm if and only if its symbol is an invertible homomorphism at each point and its normal operator is invertible on the corresponding spaces on G_p, for each $p \in \partial X$. As far as I am aware, no general index theorem for this algebra has been proved. However, an index theorem of Plamenevskii and Rozenblyum [94] can be regarded as a particular case of such a conjectural result.

8.4 The Laplacian

One direct consequence of the 0-geometry just discussed is that the Laplacian of a conformally compact metric, (8.14), is necessarily an element of $\text{Diff}_0^2(X)$. In fact the same is true for the operator acting on forms, provided these are rescaled. If $^0 \Lambda^k$ is the kth exterior power of $^0 T^* X$ then, for any k,

$$(8.28) \qquad \Delta \in \text{Diff}_0^2(X; {}^0 \Lambda^k).$$

The algebra of 0-pseudodifferential operators is designed so that

---

[22] Thus

$$(8.26) \qquad N(P \cdot Q) = N(P) \cdot N(Q), \quad P \in \text{Diff}_0^m(X), \; Q \in \text{Diff}_0^{m'}(X).$$

Proposition 8.1 *For any conformally compact metric, with curvature constant at infinity in the sense of (8.15), the (true) resolvent family*

$$(8.29) \qquad (\varDelta - \tau)^{-1} \in \Psi_0^{-2,0}(X), \ \tau \in \mathbb{C} \setminus [\frac{(n-1)^2}{4}, \infty)$$

depends meromorphically on τ and the spectrum of $\varDelta$ consists of the half-line $[(n-1)^2/4, \infty)$ with possibly a finite number of eigenvalues, of finite multiplicity, in $(0, (n-1)^2/4)$.

There is a similar result for the action on k-forms.[23] Under the assumption that $|dx|_\infty$ is constant[24] on the boundary, as in (8.16), there are no embedded eigenvalues[25] and the spectrum has uniform, infinite, multiplicity. More precisely the generalized eigenspaces can be parametrized much as in the Euclidean case.

Proposition 8.2 [26] *For a metric (8.16), each $0 \neq \lambda \in \mathbb{R}$ and each $f \in C^\infty(\partial X)$ there is a unique solution of the equation $(\varDelta - \lambda^2 - \frac{1}{4}(n-1)^2)u = 0$ of the form*

$$(8.31)$$
$$u = x^{i\lambda} x^{\frac{1}{2}(n-1)} f + x^{-i\lambda} x^{\frac{1}{2}(n-1)} f_- + u', \ f_- \in C^\infty(\partial X), \ u' \in L^2(X).$$

As usual the uniqueness of the solution in (8.30) leads to the scattering matrix

$$(8.32) \qquad A(\lambda) : C^\infty(\partial X) \longrightarrow C^\infty(\partial X).$$

In this case $A(\lambda) \in \Psi^0(\partial X)$ is an invertible pseudodifferential operator, on the compact manifold ∂X.[27] Even in the case of hyperbolic space itself the scattering matrix is not 'trivial.'

The key to understanding the behaviour of the scattering matrix, in particular why it is a pseudodifferential operator, lies in the behaviour

---

[23] See the paper of Mazzeo [58].
[24] These results can be extended to the case of a 'variable curvature at infinity' to some degree, see [58].
[25] See R. Mazzeo [59].
[26] As in the case discussed above this can be refined to give a functional parametrization of the continuous spectrum. Namely for given $f \in C^{-\infty}(\partial X)$ there is still a unique generalized eigenfunction of the form (8.31), where $f_- \in C^{-\infty}(\partial X)$ and the error term u' is permitted to be in the Sobolev space

$$(8.30) \qquad H_0^{-\infty,0}(X) = \text{Diff}_0^*(X) \cdot L^2(X), \ L^2(X) = H_0^{0,0}(X)$$

being the metric L^2 space. Furthermore these solutions exhaust the null space of $\varDelta - \lambda^2 - \frac{1}{4}(n-1)^2$ acting on $C^{-\infty}(X) = H_0^{\infty,-\infty}(X)$.
[27] It is really of 'complex order' $-2i\lambda$.

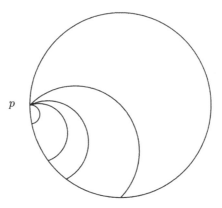

Fig. 13. Geodesics for a conformally compact metric.

of the geodesics of the conformally compact metric. It is quite easy to see that for every $p \in \partial X$ there are many geodesics which have limit at p as $t \to -\infty$. Moreover, given a neighbourhood of the boundary, $\{x < \epsilon\}$, an infinite family of these geodesics stay within that neighbourhood and have limit point, as $t \to \infty$, also on the boundary. As $\epsilon \downarrow 0$ the end point of such geodesics necessarily approaches p. Thus the boundary relation, which for a scattering metric gives geodesic flow at length π for the boundary metric, gives the identity relation for a conformally compact metric.

I shall not discuss the Hodge cohomology of metrics of this type; this has been worked out by Mazzeo [58]; see also the work of Mazzeo and Phillips [63] where the cohomology of more general hyperbolic quotients is computed.

8.5 Analytic continuation

The family of operators $(\Delta - \lambda^2 - \frac{1}{4}(n-1)^2)^{-1}$, where Δ is the Laplacian as in Proposition 8.1, extends to be a meromorphic function of $\lambda \in \mathbb{C}$, as an operator from $\mathcal{C}_c^\infty(X^\circ))$ to $\mathcal{C}^\infty(X^\circ)$. This was shown in [61], with the caveat that the residues of the poles are not shown there to be of finite rank. The finiteness of the rank in case the metric has constant curvature near infinity is shown by Guillopé and Zworski, in [32], [31]. They also show, under the same conditions, a polynomial bound on the counting function for the poles, defined as in (4.23),

$$(8.33) \qquad N(R) \le C + CR^{n+1}.$$

This is probably not optimal. For quotients of hyperbolic space of this type the analytic continuation was also shown by Froese, Hislop and Perry in [23]; see the references there for earlier work, especially that related to hyperbolic quotients.

8.6 Finite volume quotients

The second class of metrics I wish to briefly discuss today are the hyperbolic cusp metrics, as in (8.4). One feature which immediately distinguishes these metrics from all the other complete metrics I have considered so far is that they have finite volume.[28] Indeed the most familiar example of these sorts of metrics come from finite volume quotients of hyperbolic space. Consider the quotient of hyperbolic space by the infinite cyclic group generated by the single[29] transformation

$$(8.34) \qquad x + iy \longmapsto x + i(y + 1).$$

The quotient $\mathbb{H}^2/\Gamma$ is a cylinder with metric (8.13), with $|dy|^2$ interpreted as the metric on the circle. Notice in particular that this manifold with boundary has two boundary hypersurfaces, one which is of type 0 and the other of type 'hc.' The inversion $x \mapsto 1/x$ interchanges the two.

8.7 hc-geometry

The finiteness of the volume of manifolds with hc-metrics is not the only novelty as compared to the other cases considered so far. In fact the underlying difference is that the space of vector fields of bounded length is not spanned by its smooth elements.[30] This is closely related to the fact that the boundary[31] of a finite volume quotient is typically a torus. Said a different way, the hc-structure is not a boundary fibration

---

[28] The volume form near the boundary is necessarily of the form $x^{n-2}dxdy$ where dy is a volume form on the boundary and $n = \dim X$. Since $n \ge 2$, for non-triviality, this is integrable so the volume is finite.

[29] This is a typical parabolic element of PSL(2, $\mathbb{R}$).

[30] It contains elements with unbounded coefficients of the type V/x with V a smooth vector field tangent to the boundary but which does not vanish identically after restriction to the boundary.

[31] Meaning the boundary of the appropriate compactification of the quotient.

structure in the sense of [66]. The simple remedy is to regard these
hc-metrics as conformal cusp metrics. That is, simply write (8.4) as

$$(8.35) \qquad\qquad g_{hc} = x^2 g_c$$

where g_c is a metric of the form (8.10).[32] If $\mathrm{Diff}_c^*(X)$ is defined as the
enveloping algebra of the Lie algebra $\mathcal{V}_c(X)$ of cusp vector fields[33] then
the Laplacian of an hc-metric is more singular

$$(8.36) \qquad\qquad \Delta_{hc} \in x^{-2} \mathrm{Diff}_c^2(X).$$

However, this means that the resolvent family can be written in the form

$$(8.37) \qquad (\Delta_{hc} - \tau)^{-1} = x^{-2}(x^2\Delta - x^2\tau)^{-2} \in \Psi_c^{-2,-2}(X).$$

Of course I have not defined the algebra, $\Psi_c^{m,k}(X)$, of cusp pseudodifferential operators; however, I have already pointed out, in Footnote 6,
the relationship between the cusp- and b-algebras. Thus it is not too
hard to make sense of (8.37). What I really want to indicate here is that
it should lead one to expect, in general for an hc-metric, that the full
spectrum should behave like the spectrum of a b-Laplacian near 0. The
only real difference is that the spectrum will not in general start at 0.

8.8 Spectrum

The spectral theory of the Laplacian for metrics of this type has been
discussed in some detail by Müller [83], following much work, by many
authors and over a long period of time, on the finite volume quotients
of hyperbolic space.

Proposition 8.3 *The spectrum of the Laplacian of an hc-metric*[34] *con-*
sists of continuous spectrum of finite multiplicity in $[\frac{1}{4}(n-1)^2, \infty)$ with
the possibility of embedded L^2 (pure point) spectrum and also of a fi-
nite number of L^2 eigenvalues in $[0, \frac{1}{4}(n-1)^2]$. For each $0 \neq \lambda \in \mathbb{R}$

---

[32] This is one reason for singling out the 'cusp' metrics, but not the only one.
[33] I have sloughed over some subtleties here. In distinction to the earlier cases, the
cusp structure is not 'canonical.' That is, the Lie algebra $\mathcal{V}_c(X)$ is not invariant
under all diffeomorphisms of X. In fact it is equivalent to specifying a boundary
defining function up to quadratic terms or a constant multiple, i.e. it is equivalent to a choice of trivialization of the conormal bundle to the boundary, up to
a constant factor. Under a general diffeomorphism of X one cusp structure is
transformed into another. As in the other cases there is a well defined tangent
bundle corresponding to each cusp structure.
[34] Of product type.

and each element, f, of the null space of Δ_∂[35] there is a solution of
$(\Delta - \frac{1}{4}(n-1)^2 - \lambda^2)u = 0$ *of the form*

(8.38) $u = x^{i\lambda}x^{1-n}f + x^{-i\lambda}x^{1-n}f_- + u', \; u' \in L^2_{hc}(X)$

which is unique up to addition of an L^2 eigenfunction with eigenvalue
$\frac{1}{4}(n-1)^2 + \lambda^2$.

The scattering matrix in this case is a matrix!Namely it is a λ-dependent linear operator on the finite-dimensional space consisting of the null space of the boundary Laplacian. This is again easy to relate to the behaviour of the geodesics. Indeed there is a neighbourhood of the boundary which contains no complete geodesic. Again the resolvent has a meromorphic continuation to the λ plane. For a discussion of the poles of the continuation see the papers of Guillopé and Zworski [32] and Froese and Zworski [24]. The trace formula, giving the regularized trace of the wave group in terms of the eigenvalues and the scattering matrix is well-known in the case of hyperbolic quotients, as Selberg's trace formula, and extends to the general case in a similar form.

---

[35] Note that the boundary may consist of several components. The null space of Δ_∂ on each component just consists of the constants, but these can differ from component to component.

References

[1] S. Agmon, *Spectral theory of Schrödinger operators on Euclidean and non-Euclidean spaces*, Comm. Pure Appl. Math. **39** (1986), 3–16.

[2] N. Anghel, L^2-*index formula for perturbed Dirac operators*, Comm. Math. Phys. **128** (1990), 77–97.

[3] _____, *On the index of Callias-type operators*, Geom. and Funct. Anal. **3** (1993), 431–438.

[4] M.F. Atiyah, *K-theory*, Benjamin, New York, 1967.

[5] M.F. Atiyah and I.M. Singer, *The index of elliptic operators, III*, Ann. of Math. **87** (1968), 546–604.

[6] C. Bardos, J.-C. Guillot, and J. Ralston, *La relation de Poisson pour l'équation des ondes dans un ouvert non borné*, Comm P.D.E. **7** (1982), 905–958.

[7] C. Bardos, G. Lebeau, and J. Rauch, *Scattering frequencies and Gevrey 3 singularities*, Invent. Math. **90** (1987), 77–1144.

[8] A. Sá Barreto and M. Zworski, *Existence of resonances in three dimensions*, Preprint, Sept 1994.

[9] R. Beals and R.R. Coifman, *Multi-dimensional inverse scattering and nonlinear partial differential equations*, Proc. Symp. Pure Math. **43** (1985), 45–70.

[10] J.-M. Bismut and J. Cheeger, *Families index for manifolds with boundary, superconnections and cones*, Invent. Math. **89** (1990), 91–151.

[11] B.H. Bowditch, *Geometrical finiteness for hyperbolic groups*, J. Funct. Anal., To appear.

[12] J. Brüning and V.W. Guillemin (Editors), *Fourier integral operators*, Springer-Verlag, Berlin, Heidelberg, New York, Tokyo, 1994.

[13] A.P. Calderón, *On an inverse boundary value problem*, Seminar on Numerical Analysis and its Applications to Continuum Physics (Rio de Janeiro), vol. 90, Soc. Brasileira de Matemàtica, pp. 65–73.

[14] C. Callias, *Axial anomalies and the index theorem on open spaces*, Comm. Math. Phys. **62** (1978), 213–234.

[15] T. Christiansen, *Scattering theory for manifolds with asymptotically cylindrical ends*, J. Funct. Anal., To appear.

[16] T. Christiansen and M. Zworski, *Spectral asymptotics for manifolds with cylindrical ends*, Preprint, 1994.

[17] H.L. Cycon, R.G. Froese, W. Kirsch, and B. Simon, *Schrödinger operators*, Springer-Verlag, 1987.

[18] H. Donnelly, *Eigenvalue estimates for certain noncompact manifolds*, Michigan Math. J. **31** (1984), 349–357.

[19] J.J. Duistermaat and V.W. Guillemin, *The spectrum of positive elliptic operators and periodic geodesics*, Invent. Math. **29** (1975), 39–79.

[20] A.K. Erkip and E. Schrohe, *Normal solvability of boundary value problems on asymptotically flat manifolds*, J. Funct. Anal. **109** (1992), 22–51.

[21] G. Eskin and J. Ralston, *The inverse backscattering problem in three dimensions*, Comm. Math. Phys. **124** (1989), 169–215.

[22] F.G. Friedlander, *The wave equation on a curved space-time*, Cambridge University Press, Cambridge, London, New York, Melbourne, 1975.

[23] R. Froese, P. Hislop, and P. Perry, *A Mourre estimates and related bounds for hyperbolic manifolds with cusps of non maximal rank*, J. Funct. Anal. **98** (1991), 292–310.

[24] R. Froese and M. Zworski, *Finite volume surfaces with resonances far from the unitarity axis*, Int. Math. Research Notices **10** (1993), 275–277.

[25] I.M. Gel'fand and B.M. Levitan, *On the determination of a differential equation from its spectral function*, Izv. Akad. Nauk SSSR **15** (1951), 309–360.

[26] C. Gérard, *Asymptotique des pôles de la matrice de scattering pour deux obstacles strictement convexes*, Bull. Soc. Math. France **116** (1988).

[27] I. Gohberg and M. Krein, *Introduction to the theory of non-self-adjoint operators*, A.M.S., Providence, R.I., 1969.

[28] I.C. Gohberg and E.I Sigal, *An operator generalization of the loga-rithmic residue theorem and the theorem of Rouché*, Math. USSR Sbornik **13** (1971), no. 4, 603–624, Russian Volume 126.

[29] V.W. Guillemin, *Sojourn times and asymptotic properties of the scattering matrix*, RIMS Kyoto Univ. **12** (1977), 69–88.

[30] V.W. Guillemin and S. Sternberg, *Geometrical asymptotics*, AMS Surveys, vol. 14, AMS, Providence, R.I., 1977.

[31] L. Guillopé and M. Zworski, *Polynomial bounds on the number of resonances for some complete spaces of constant negative curvature near infinity*, Preprint, 1994.

[32] ———, *Upper bounds on the number of resonances for non-compact Riemann surfaces*, Preprint, 1994.

[33] J. Hadamard, *Le problème de Cauchy et les èquatons aux dérivées partielles linéaires hyperboliques*, Hermann, Paris, 1932.

[34] T. Hargé and G. Lebeau, *Diffraction par un convexe*, Invent. Math. **118** (1994), 161–196.

[35] A. Hassell, R. Mazzeo, and R. Melrose, *Analytic surgery and the eta invariant, II*, Preprint, 1994.

[36] S. Helgason, *The Radon transform on Euclidean spaces, compact two-point homogeneous spaces and Grassmann manifolds*, Acta Math. **113** (1965), 153–180, See also [81].

[37] G.M. Henkin and R.G. Novikov, *The $\bar{\partial}$-equation in the multidi-mensional inverse scattering problem*, Russion Math. Surveys **42** (1987), 109–180.

[38] L. Hörmander, *Pseudo-differential operators*, Comm. Pure Appl. Math. **18** (1965), 501–517.

[39] ———, *The spectral function of an elliptic operator*, Acta Math. **121** (1968), 193–218, See also [12].

[40] ———, *Fourier integral operators, I*, Acta Math. **127** (1971), 79–183, See also [12].

[41] ———, *The Weyl calculus of pseudo-differential operators*, Comm. Pure Appl. Math. **32** (1979), 359–443.

[42] ———, *The analysis of linear partial differential operators*, vol. 1, Springer-Verlag, Berlin, Heidelberg, New York, Tokyo, 1983.

[43] ———, *The analysis of linear partial differential operators*, vol. 4, Springer-Verlag, Berlin, Heidelberg, New York, Tokyo, 1985.

[44] ———, *The analysis of linear partial differential operators*, vol. 3, Springer-Verlag, Berlin, Heidelberg, New York, Tokyo, 1985.

[45] M. Ikawa, *On the poles of the scattering matrix for two strictly convex obstacles*, J. Math. Kyoto Univ. **23** (1983), 127–194.

[46] A. Intissar, *A polynomial bound on the number on the scattering poles for a potential in even dimensions*, Comm. P.D.E. **11** (1986), 367–396.

[47] F. John, *Plane waves and spherical means applied to partial differential equations*, Wiley (Interscience), 1955.

[48] T. Kato, *Wave operators and unitary equivalence*, Pacific J. Math. **15** (1965), 171–180.

[49] J.J. Kohn and L. Nirenberg, *On the algebra of pseudo-differential operators*, Comm. Pure Appl. Math. **18** (1965), 269–305.

[50] V.A. Kondrat'ev, *Boundary value problems for elliptic equations in domains with conical points*, Trudy Mosk. Math. Obsc. **16** (1967), 209–292.

[51] P.D. Lax and R.S. Phillips, *Scattering theory*, Academic Press, New York, 1967, Revised edition, 1989.

[52] _____, *Decaying modes for the wave equation in the exterior of an obstacle*, Comm. Pure Appl. Math. **22** (1969), 737–787.

[53] _____, *A logarithmic bound on the location of the poles of the scattering operator*, Arch. Rat. Mech. Anal. **40** (1971), 268–280.

[54] _____, *Scattering theory for automorphic forms*, Ann. of Math. Studies, vol. 87, Princeton University Press, 1976.

[55] G. Lebeau, *Regularité Gevrey 3 pour la diffraction*, Comm. in P.D.E. **9** (1984), 1437–1494.

[56] A. Majda, *High frequency asymptotics for the scattering matrix and the inverse problem of acoustical scattering*, Comm. Pure Appl. Math. **29** (1976), 261–291.

[57] _____, *A representation for the scattering operator and the inverse problem for arbitrary bodies*, Comm. Pure Appl. Math. **30** (1977), 167–196.

[58] R. Mazzeo, *The Hodge cohomology of a conformally compact metric*, J. Diff. Geom. **28** (1988), 309–339.

[59] _____, *Unique continuation at infinity and embedded eigenvalues for asymptotically hyperbolic manifolds*, Amer. J. Math. **113** (1991), 25–46.

[60] R. Mazzeo and R. Melrose, *Analytic surgery and the eta invariant*, Geom. and Func. Anal., To appear.

[61] _____, *Meromorphic extension of the resolvent on complete spaces with asymptotically constant negative curvature*, J. Funct. Anal. **108** (1987), 260–310.

[62] _____, *The adiabatic limit, Hodge cohomology and Leray's spectral sequence*, J. Diff. Geom. **31** (1990), 185–213.

[63] R. Mazzeo and R. Phillips, *Hodge theory on hyperbolic manifolds*, Duke Math. J. **60** (1990), 509–559.

[64] H. McKean and I.M. Singer, *Curvature and the eigenvalues of the Laplacian*, J. Diff. Geom. **1** (1967), 43–69.

[65] A. Melin, *Inverse scattering problem for a quantum mechanical two-body system*, Proceedings of the twenty-first Nordic congress of mathematicians, Lecture notes in pure and applied mathematics, vol. 156, Marcel Dekker, 1994, pp. 247–261.

[66] R.B. Melrose, *Differential analysis on manifolds with corners*, In preparation.

[67] _____, *Microlocal parametrices for diffractive boundary value problems*, Duke Math. J. **42** (1975), 605–635.

[68] _____, *Transformation of boundary problems*, Acta Math. **147** (1981), 149–236.

[69] _____, *Scattering theory and the trace of the wave group*, J. Funct. Anal. **45** (1982), 429–440.

[70] _____, *Polynomial bound on the number of scattering poles*, J. Funct. Anal. **53** (1983), 287–303.

[71] _____, *Polynomial bounds on the distribution of poles in scattering by an obstacle*, Journeés "Equations aux Dérivées Partielle", Saint-Jean-des-Monts, 1984.

[72] _____, *Weyl asymptotics for the phase in obstacle scattering*, Comm. in P.D.E. **13** (1988), 1421–1439.

[73] _____, *The Atiyah-Patodi-Singer index theorem*, A K Peters, Wellesley, MA, 1993.

[74] _____, *Spectral and scattering theory for the Laplacian on asymptotically Euclidian spaces*, Spectral and scattering theory (M. Ikawa, ed.), Marcel Dekker, 1994.

[75] R.B. Melrose and G.A. Mendoza, *Elliptic pseudodifferential operators of totally characteristic type*, MSRI Preprint, 1983.

[76] R.B. Melrose and P. Piazza, *Analytic K-theory for manifolds with corners*, Adv. Math. **92** (1992), 1–27.

[77] _____, *Families of Dirac operators, boundaries and the b-calculus*, Preprint, 1993.

[78] _____, *An index theorem for families of Dirac operators on odd-dimensional manifolds with boundary*, Preprint, 1994.

[79] R.B. Melrose and J. Sjöstrand, *Singularities in boundary value problems I*, Comm. Pure Appl. Math. **31** (1978), 593–617.

[80] _____, *Singularities in boundary value problems II*, Comm. Pure Appl. Math. **35** (1982), 129–168.

[81] R.B. Melrose and G. Uhlmann, *An introduction to microlocal analysis and scattering theory*, Preliminary version, 1994.

[82] R.B. Melrose and M. Zworski, *Scattering metrics and geodesic flow at infinity*, Preprint, 1995.

[83] W. Müller, *Manifolds with cusps of rank one*, Lecture Notes in Math., vol. 1244, Springer Verlag, Berlin, Heidelberg, New York, 1988.

[84] A.I. Nachman, *Reconstruction from boundary measurements*, Ann. Math. (1988), 531–587.

[85] R.G. Novikov, *Multidimensional inverse spectral problems for the equation* $-\Delta\psi + (v(x) - E)\psi = 0$, Funct. Anal. and Appl. **22** (1988), 263–272.

[86] _____, *The inverse scattering problem on a fixed energy level for two-dimensional Schrödinger operator*, J. Funct. Anal. **103** (1992), 409–463.

[87] _____, *Reconstruction of an exponentially decreasing potential for the three-dimensional Schrödinger equation through the scattering amplitude at fixed energy*, C. R. Acad. Sci. Paris **316** (1993), 657–662.

[88] C. Parenti, *Operati pseudodifferenziali in* $\mathbb{R}^n$ *e applicazioni*, Annali Mat. Pura et Appl. **93** (1972), 359–389.

[89] V. Petkov and G. Vodev, *Upper bounds on the number of scattering poles and the Lax-Phillips conjecture*, Asympt. Anal. **7** (1993), 87–104.

[90] R.S. Phillips, *Scattering theory for the wave equation with a short range perturbation, I*, Indiana Univ. Math. J. **31** (1982), 609–639.

[91] _____, *Scattering theory for the wave equation with a short range perturbation, II*, Indiana Univ. Math. J. **33** (1984), 831–846.

[92] P. Piazza, *On the index of elliptic operators on manifolds with boundary*, J. Funct. Anal. **117** (1993), 308–359.

[93] B.A. Plamenevskii, *Algebras of pseudo-differential operators*, Nauka, Moscow, 1986, (Russian).

[94] B.A. Plamenevskii and G.V. Rozenblyum, *Pseudodifferential operators with discontinuous symbols: K-theory and index formula*, Funk. Anal. Appl. (Russian) **26** (1991), no. 4, 45–56.

[95] G. Popov, *Some estimates of Green's function in the shadow*, Osaka J. Math. **24** (1987), 1–12.

[96] J. Råde, *An* L^2 *index theorem for Dirac operators on complete odd-dimensional manifolds*, Preprint, 1993.

[97] J. Rauch, *Illumination of bounded domains*, Amer. Math. Monthly **85** (1978), no. 5, 359.

[98] M. Reed and B. Simon, *Methods of modern mathematical physics*, vol. III, Academic Press, 1979.

[99] J.R. Ringrose, *Compact non-self-adjoint operators*, Van Nostrand Reinhold Company, 1971.

[100] D. Robert, *Relative time-delay for perturbations of elliptic operators and semiclassical asymptotics*, Jour. Funct. Anal., To appear.

[101] _____, *Relative time delay and trace formula for long range perturbation of Laplace's operator*, Operator theory; Advances and Applications, vol. 57, Birkäuser Verlag, Basel.

[102] B.-W. Schulze, *Corner Mellin operators and reductions of order with parameters.*, Ann. Sc. Norm. Sup., Pisa **16** (1989), 1–81.

[103] L. Schwartz, *Théorie des distributions I, II*, Hermann, Paris, 1950-51.

[104] R.T. Seeley, *Complex powers of elliptic operators*, Proc. Symp. Pure Math. **10** (1967), 288–307.

[105] _____, *The resolvent of an elliptic boundary problem*, Amer. J. Math. **91** (1969), 889–919.

[106] M.A. Shubin, *Pseudodifferential operators on* $\mathbb{R}^n$, Sov. Math. Dokl. **12** (1971), 147–151.

[107] _____, *Spectral theory of elliptic operators on non-compact manifolds*, Summer school on semiclassical methods, Nantes.

[108] J. Sjöstrand and M. Zworski, *Lower bounds on the number of scattering poles II*, J. Funct. Anal., To appear.

[109] _____, *Complex scaling and the distribution of scattering poles*, J. Amer. Math. Soc. **4** (1991), 729–769.

[110] _____, *Estimates on the number of scattering poles near the real axis for scattering by strictly convex obstacles*, Ann. l'Inst. Fourier **43** (1993), 769–790.

[111] J. Sylvester and G. Uhlmann, *A global uniqueness theorem for an inverse boundary value problem*, Ann. of Math. **125** (1987), 153–169.

[112] M.E. Taylor, *Grazing rays and reflection of singularities to wave equations*, Comm. Pure Appl. Math. **29** (1978), 1–38.

[113] G. Uhlmann, *Inverse boundary value problems and applications*, Asterisque **207** (1992), 153–211.

[114] A. Vasy, *Scattering poles for positive potentials*, preprint, 1994.

[115] G. Vodev, *On the distribution of scattering poles for perturbations of the Laplacian*, Ann. Inst. Fourier **40** (1992), 625–635.

[116] ———, *Sharp polynomial bounds on the number of scattering poles for perturbations of the Laplacian*, Comm. Math. Phys. **146** (1992), 205–216.

[117] ———, *Sharp bounds on the number of scattering poles in even dimensional spaces*, Preprint, 1994.

[118] D. Yafaev, *Mathematical scattering theory*, A.M.S., Providence, R.I., 1992.

[119] M. Zworski, *Sharp polynomial bounds on the number of scattering poles*, Duke Math. Jour. **59** (1989), no. 2, 311–323.

[120] ———, *Sharp polynomial bounds on the number of scattering poles of radial potentials*, J. Funct. Anal. **82** (1989), no. 2, 370–403.

[121] ———, *Counting scattering poles*, Spectral and scattering theory (M. Ikawa, ed.), Marcel Dekker.

Index

Printed in the United States
By Bookmasters